[英国] 伊恩·汤普森 著　安聪 译

景观设计学

牛津通识读本·

Landscape Architecture

A Very Short Introduction

译林出版社

图书在版编目（CIP）数据

景观设计学 /（英）伊恩·汤普森（Ian Thompson）著；
安聪译. —南京：译林出版社，2022.2
（牛津通识读本）
书名原文：Landscape Architecture: A Very Short Introduction
ISBN 978-7-5447-8870-0

Ⅰ.①景… Ⅱ.①伊… ②安… Ⅲ.①景观设计
Ⅳ.①TU983

中国版本图书馆 CIP 数据核字（2021）第 204822 号

著作权合同登记号 图字：10-2018-429 号

景观设计学 ［英国］伊恩·汤普森 / 著 安 聪 / 译

责任编辑 杨欣露
装帧设计 景秋萍
校　　对 王　敏
责任印制 董　虎

原文出版 Oxford University Press, 2014
出版发行 译林出版社
地　　址 南京市湖南路 1 号 A 楼
邮　　箱 yilin@yilin.com
网　　址 www.yilin.com
市场热线 025-86633278
排　　版 南京展望文化发展有限公司
印　　刷 江苏扬中印刷有限公司
开　　本 635 毫米 ×889 毫米 1/16
印　　张 17.75
插　　页 4
版　　次 2022 年 2 月第 1 版
印　　次 2022 年 2 月第 1 次印刷
书　　号 ISBN 978-7-5447-8870-0
定　　价 39.00 元

序　言

俞孔坚

当下，全球气候变化日益严峻，城镇蔓延，洪涝风险和水土污染威胁人类生存，生物多样性日益丧失，整体人居环境的不确定性不断加剧。试问世界上还有哪些学科与职业比规划和设计安全、健康、美丽的人类家园更重要？“牛津通识读本”《景观设计学》介绍的正是这样一门古老而崭新的学科，它以人与自然和谐共生为目标，通过综合协调人与自然、当代人的活动与历史文化遗产的关系，将科学与艺术相结合，在满足人类物质欲望的同时，追求人类精神与审美富足的美好家园。本书所隶属的“牛津通识读本”系列是牛津大学出版社的重点项目，英文原版自1995年起陆续面世以来，在全球范围内已被译成近五十种文字。丛书主题广泛，涵盖哲学、科学、艺术、文化、历史、经济等领域，作者多为国外大学或研究机构的知名学者，被誉为真正的“大家小书”。

要读懂这本小书，不仅需要理解文艺复兴以来西方现代学科的发展和分类规律，也需要理解从农业社会的自给自足和经验主义，到工业社会中社会化的职业分工及科学体系的建立，再到当代系统科学尤其是生态系统科学的发展对学科和专业发展的影响；还需要理解农业时代的贵族小范围的造园（gardening）如何走向工业化时代面向大众开放的风景造园（landscape gardening），再到城市和大地上景观的科学规划与设计（landscape architecture）。由于中国的现代学科体系源自现代西方，所以，首先需要搞清楚的是中西文语境下的一些基本概念，特别是本书中的一些关键专业术语和概念，包括①：

1. Landscape　景观（当landscape被用来描绘自然风景画时，则翻译为风景，如landscape painting和landscape art，尤指宽广视野下描绘的山水、森林等自然景物）

2. Scenery　风景，景致

3. Garden　花园

4. Gardening　造园

5. Horticulture　园艺

6. Landscape garden　风景园，风景园林［早期的西方花园（garden）源于园艺，主要用植物来营造。当山水风景被引入园林，特别是受到中国的文人山水园林中假山水的影响后，英国人便在花园前加了landscape一词来描述有山水和自然风景的花园，而后又融入了源于风景画的画意园林，并将花园外的风景与花园内的园艺融为一体］

7. Landscape gardening　风景造园，风景园林营造

① 俞孔坚、李迪华，《景观设计：专业　学科与教育》，中国建筑工业出版社，2003。

8. Landscape architecture　景观设计学［包含景观规划（landscape planning）和景观设计（landscape design）两个分支。国内往往译为风景园林，与landscape gardening相混淆。Architecture的实质是设计学，如computer architecture］

9. Landscape design　景观设计（即具体的设计，指明该如何做，更多取决于设计师的科学和艺术修养）

10. Landscape planning　景观规划（即划定边界，明确在什么地方干什么事，取决于科学分析和决策过程）

11. Landscape urbanism　景观都市主义（是景观而不是建筑决定了城市的形态与布局）

12. Landscape ecology　景观生态学（研究生态系统之间的空间和生态流的关系与变化）

13. Horticulturist　园艺师

14. Gardener　造园师

15. Garden design　花园设计

16. Garden designer　花园设计者

17. Landscape gardener　风景造园师

18. Landscape designer　景观设计者

19. Landscape architect　职业景观规划与设计师（一般限于注册的职业设计师）

关于景观的含义

景观（landscape），无论在西方还是在中国都是一个美丽而难以说清的概念。地理学家把景观视作一个科学名词，定义为一种地表景象，或综合自然地理区，或是用作某种类型的地表景物的通称，如城市景观、草原景观、森林景观等[①]；艺术家把景观作为表现与再现的对象，等同于风景；建筑师把景观作为建筑物的配景或背景；生态学家把景观定义为生态系统或生态系统的集合[②]；旅游学家把景观当作资源；更常见的是景观被城市美化运动者和开发商等同于城市的街景立面、霓虹灯，以及房地产中的园林绿化和小品、喷泉叠水；更文学和宽泛的定义则是"能用一个画面来展示，能在某一视点上可以全览的景象"，尤其是自然景象。但哪怕是对同一景象，不同的人也会有很不同的理解，正如梅尼（Meinig）所认为的"同一景象的十个版本"，即景观是人所向往的自然，景观是人类的栖居地，景观是人造的工艺品，景观是需要科学分析方能被理解的物质系统，景观是有待解决的问题，景观是可以带来财富的资源，景观是反映社会伦理、道德和价值观念的意识形态，景观是历史，景观是场所，景观是美[③]。

作为景观设计对象，本书所强调的景观是指土地及土地上的空间和物体所构成的综合体。它是复杂的自然过程和人类活动在大地上的烙印。景观是多种功能（过程）的载体，因而可被理解和表现为[④]：

1. 风景（视觉审美过程的对象）

2. 栖居地（人类生活其中的空间和环境）

3. 生态系统（一个具有结构和功能、具有内在和外在联系的

① 辞海编辑委员会，《辞海》，上海辞书出版社，1995。

② Zev Naveh, et al., *Landscape Ecology: Theory and Application*, Springer, 1984; R. T. T. Forman, et al., *Landscape Ecology*, John Wiley, 1986.

③ D. W. Meinig, "The Beholding Eye: Ten Versions of the Same Scene", *Landscape Architecture*, 1976(1). pp. 47—53.

④ 俞孔坚，《景观的含义》，刊于《时代建筑》，2002年第1卷，第14—17页。

有生命的系统）

4. 符号（一种记载人类的过去、表达希望与理想的语言和精神空间）

既然景观是一种综合体，我们就要考虑如何系统、全面地设计美的景物，如何设计人与自然、人与人和谐的社区，如何设计健康的生态系统，如何体现文化含义。因此就出现了一门叫景观设计学的学科，它的前身之一是为美的目的设计和建造风景与园林，但更久远的根在于人类适应自然和改造自然的一切活动所积累的生存智慧，包括开垦农田、灌溉、种植和家园的设计及营造。19世纪末的工业化进程促进了景观设计学和职业设计师的产生，工业化的最大特点是职业的社会化，因此就产生了系统地整合这门学科的必要。1900年，哈佛大学出现了景观设计学课程，景观设计学才真正成为一门学科。不过在此之前，这一学科可以说已有四五十年的实践经验，因为创始人奥姆斯特德（Olmsted）从19世纪60年代就已开始景观设计。23年后，从景观设计学科中又分出了城市规划设计学科。这门学科是对土地的全面设计，其核心是人如何利用土地、协调人与自然的关系，是一门用人类积累的科学和艺术智慧来设计人类美好家园的应用性、实践性学科。

关于景观设计学

景观设计学是关于景观的分析、规划布局、设计、改造、管理、保护和修复的科学与艺术。

作为一门建立在广泛的自然科学和人文与艺术学科基础上的应用学科，景观设计学尤其强调土地的设计，即通过对有关土地及一切人类户外空间的问题进行科学理性的分析，科学并艺术地设计问题的解决方案和解决途径，同时监理设计的实现。

根据解决问题的性质、内容和尺度的不同，景观设计学包含两个专业方向，即景观规划和景观设计。前者是指在较大尺度范围内，基于对自然和人文的认识，协调人与自然关系的过程，具体说是为某些使用目的安排最合适的地方和在特定的地方安排最恰当的空间与土地利用；而对某个特定用途的地方的设计就是景观设计。

景观设计学与建筑学、城市规划、环境艺术、市政工程设计等应用学科有紧密的联系，而景观设计学所关注的问题是土地和人类户外空间的问题（仅这一点就有别于建筑学）。它与现代意义上的城市规划的主要区别在于，景观设计学是对物质空间的规划和设计，包括城市与区域的物质空间规划设计，而城市规划更主要关注社会经济和城市总体发展计划。然而，中国目前的城市规划专业仍在主要承担城市的物质空间规划设计，这是中国景观设计和城市规划等学科发展滞后的结果。只有同时掌握关于自然系统和社会系统双方面知识、懂得如何协调人与自然关系、既能理性掌握科学知识又具有艺术和审美修养的景观设计师，才有可能设计人与自然和谐共生的美好城市。

与市政工程设计不同，景观设计学更善于综合地、多目标地解决问题，而不是目标单一地解决某个工程问题。当然，综合解决问题的过程有赖于各个市政工程设计专业的参与。

与环境艺术（甚至大地艺术）的主要区别在于，景观设计学注重用综合的途径解决问题，关注对一个物质空间的整体设计，解决问题的途径建立在科学理性的分析基础上，而不仅仅依赖设计师的艺术灵感和艺术创造。

景观设计师也有别于生态修复与环境保护等专业，因为对景观设计师来说，美和艺术是生态系统与环境中不可或缺的特征。

关于景观设计师

景观设计师是以对景观的规划设计为职业的专业人员，他以实现建筑、城市和人的一切活动与充满生命的地球和谐相处为终身目标①。

景观设计师的称谓由美国景观设计之父奥姆斯特德于1858年非正式使用，1863年被正式作为职业称号②。奥姆斯特德坚持用景观设计师，而不用在当时盛行的风景造园师，不仅仅是职业称谓上的创新，也是对该职业内涵和外延的一次意义深远的扩充与革新。

景观设计师有别于传统造园师和园丁（gardener，对应gardening）、风景花园师（或称风景造园师，landscape gardener，对应landscape gardening）的根本之处在于，景观设计职业是大工业、城市化和社会化背景下的产物，是在现代科学与技术（而不仅仅是经验）的基础上发展出来的；景观设计师要处理的对象是土地综合体的复杂的综合问题，绝不只是某个层面的问题（如视觉审美意义上的风景问题）；景观设计师所面临的问题是土地、人类、城市、一切生命的安全与健康及可持续的问题，而非个人的情趣使然。他是以土地的名义、以人类和其他生命的名

① 斯塔克、西蒙兹，《景观设计学》，朱强、俞孔坚等译，中国建筑工业出版社，2000。

② Norman T. Newton, *Design on the Land: The Development of Landscape Architecture*, The Belknap Press of Harvard University, 1971.

义、以人类历史与文化遗产保护的名义，来监护、合理地利用、设计脚下的土地及土地上的空间和物体。

关于景观设计专业的发展

与建筑学一样，景观设计职业先于景观设计学形成，在大量景观设计实践的基础上，发展并完善了景观设计的理论、方法和技术，这便是景观设计学。

农业时代中西方文化里的造园艺术、前科学时代的地理思想和占地术（在中国被称为“风水”）、农业及园艺技术、不同尺度上的水利和交通工程经验、风景审美艺术、居住及城市营建技术和思想等等，都是宝贵的技术与文化遗产，是现代意义上的景观设计学的创新与发展源泉。但是，景观设计学绝不能等同于已有了约定俗成的内涵与外延的造园艺术（或园林艺术），也不能等同于风景园林艺术。

中国是世界文明古国之一，有着非常悠久的古代造园历史，也有非常精湛的传统园林艺术。同时，我们也需要认识到传统文化遗产与现代学科体系之间有着本质差别。正如算术之于数学、中国的针灸之于现代医学不能同日而语一样，任何一种源于农业时代的经验技艺，都必须经历一个用现代科学技术和理论方法进行脱胎换骨的过程，才能更好地解决大工业时代的问题，特别是城镇化带来的人地关系问题。园林艺术也是如此。早在1858年，奥姆斯特德就认识到了这一点，因此坚持将自己所从事的职业称为landscape architecture，而非当时普遍采用的landscape gardening，为景观设计专业和学科的发展开辟了一个广阔的空间，影响延续了一百六十多年。

中国后工业景观设计与上海后滩公园

随着城市化在中国的快速推进，中国的人居环境和自然生态都发生了翻天覆地的巨变，曾经优秀的传统中国园林面临了不可逾越的挑战。由于现代中国曾经历不堪回首的、与世界景观设计学科发展潮流的长期隔绝，现代学科意义上的中国景观设计学科和职业发展迟缓，错过了生机勃勃的国际现代主义时期。当20世纪80年代开始与外界发生交流和碰撞时，中国学界和业界都普遍对当代景观设计感到陌生、不解甚至抵触。虽然如此，当代景观设计与城镇化相伴共生而发展，面对迫切需要解决的人居环境和全球性气候变化挑战，得益于五千年文明的传统生态智慧，中国的景观设计在近二十年来获得了长足的发展，尤其是近十年来国家对生态文明和美丽中国建设的憧憬，对诸如国土生态规划和生态修复、黑臭河治理、城市内涝治理、乡村振兴、旧城改造、文化遗产保护等专业性工作的迫切需求，使带着明显的后工业特征的景观设计在中国大地上孕育而生。具有鲜明中国性的理念和方法，如基于自然（nature based）的设计生态学（designed ecology）和“海绵城市”（sponge city）理论等，已经被公认为中国学者对世界景观设计界的贡献。本书介绍的唯一中国当代景观设计项目上海后滩公园，即是上述理念和方法的典型代表。

后滩公园是2010上海世博园的核心景观之一，位于黄浦江东岸与浦明路之间，南临园区新建浦明路，西至倪家浜，北望卢浦大桥，占地18公顷。场地原为钢铁厂（浦东钢铁集团）和后滩船舶修理厂所在地。2007年初，由我及土人设计（Turenscape）

团队开始设计，2010年5月正式建成并对外开放。设计团队倡导足下文化与野草之美的环境伦理和新美学思想，采用了典型的后工业景观设计手法。设计显现了场地的四层历史与文明属性：黄浦江滩的自然过程，场地的农业文明与工业文明的记忆，最重要的是后工业时代生态文明理念的展望和具体实践。最终在垃圾遍地、污染严重的原工业棕地上，建成了具有水体净化和雨洪调蓄、生物生产、生物多样性保育、审美启智等综合生态服务功能的城市公园。

作为工业时代生态文明的展望和实践，公园的核心是一条带状的、具有水净化功能的人工湿地系统，它将来自黄浦江的劣Ⅴ类水，通过沉淀池、叠瀑墙、梯田、不同深度和不同群落的湿地净化区，经过长达1.7公里的流程净化成为Ⅲ类水。经过十多年的长期观察，证明了加强型人工湿地的日净化量为2 400吨，从而建立了一个可以复制的水系统的生态净化模式。设计还充分利用旧材料，倡导节约造价、低成本维护等生态理念。后滩公园深情地回望农业和工业文明的过去，并憧憬生态文明的未来，放声讴歌生态之美、丰产与健康的大脚之美、蓬勃而烂漫的野草之美的新美学观。它展示了基于自然、让自然做功的生态设计途径，为解决当下中国和世界的环境问题提供一个可以借鉴的样板，指明了建立低碳和负碳城市的一个具体路径[①]。

结　语

需要提示读者的是，奥姆斯特德给景观设计的专业和学科

① 俞孔坚，《城市景观作为生命系统——2010年上海世博会后滩公园》，刊于《建筑学报》，2010年第7卷，第30—35页。

定义的空间也绝不应是未来景观设计学科发展的界限，沿用了一百六十多年的景观设计学的名称及其内涵和外延的认知界定都已经面临巨大挑战。以对人类美好家园的规划设计和营造为核心的科学与艺术可能需要有更合适的学科称谓来涵盖。事实上，早在20世纪60年代，另一位美国景观设计学科的领袖人物麦克哈格（McHarg）就是针对当时景观设计学科无能应对城市问题、土地利用及环境问题的挑战，扛起了生态规划的大旗，使景观设计学科再次走到拯救城市、拯救人类和地球的前沿，并提出了景观生态规划（landscape ecological planning），简称生态规划（ecological planning）。半个世纪过去了，全球气候变化的加剧、大规模的城镇化、严峻的生态环境恶化、物种的大量灭绝，以及数字技术与人工智能的飞速发展，都将使人类美好家园的设计和营造面临新的问题与挑战。可持续理论、生态科学、信息技术、现代艺术理论和思潮都将为新的问题与挑战提供新的解决途径及对策。学界在重新定义这样一门学科的内涵和外延甚至名称方面，都有了新的探索，新的学科发展更是把解决城市问题、应对全球气候变化和修复全球生态系统作为重要的研究和从业内容，因而有了景观都市主义、生态都市主义（ecological urbanism）、设计生态学、地理设计（geodesign）等。

但无论学科及其称谓如何发展，景观设计学科所包含的根本，或者说它可以被一脉相承的学科基因是不变的，那就是设计和营造美好家园、通往天地人神和谐的科学与艺术：热爱土地与自然生命的伦理（天地）、以人为本的人文关怀（人）和对待地方文化与历史的尊重（神）。

目　录

前　言

景观设计学在塑造我们大部分人的日常生活和工作场所中起到了重要的作用，它扎根于环境操控的实践中，其实践历史至少与建筑和工程一样悠久。尽管如此，在许多国家，它却没有得到广泛的认可。为什么会这样？这会是我在本书中回答的问题之一，但部分原因要归咎于“景观设计学”（landscape architecture）这个有失偏颇和具有误导性的学科名称。公众对于我们怎么被这个名称误导莫衷一是。人们常说，纽约中央公园的设计者——弗雷德里克·劳·奥姆斯特德（1822—1903）和卡尔弗特·沃克斯（1824—1895）是第一个使用“景观设计师”这一称呼的人。1858年，他们在自己的胜出作品中使用了该词。但是最近，景观史学家尼娜·安东内蒂表明，精心设计了白金汉宫花园、方案却被维多利亚女王和阿尔伯特亲王否决的设计师威廉·安德鲁斯·内斯菲尔德，早在1849年就称自己为“景观设计师”了。其他学者则认为，设计师兼园艺师安德鲁·杰克逊·唐宁（1815—1852）是“第一位景观设计师”，他是最早主张在曼哈顿建立大型公园的人之一（图1）。但毫无疑问的是，奥姆斯特德和沃克斯广为人知的成功开创了景观设计师这一职业。奥姆斯特德以其在保护自然和改善城市卫生方面的贡献而闻名，但他最伟大的遗作则是他在美国许多城市设计的公园，马萨诸塞州的波士顿、纽约州的布鲁克林和布法罗、伊利诺伊州的芝加哥、肯塔基州的路易斯维尔，以及威斯康星州的密尔沃基等地都有他的作品。受英国风景造园传统的影响，奥姆斯特德认为，在城市环境中创造田园风光可以为亟须脱离喧嚣、繁忙和压力的城市居民提供缓解之机。很明显，中央公园设计竞赛中胜出的方案“草坪计划”包含了许多别具特色的景点，如漫步区、绵羊草坪、迪恩小径和大草坪等。

尽管奥姆斯特德是景观设计学的创始人，他对这个名字却一直心存疑虑。“景观设计学这个可怜的名称一直困扰着我，”他在1865年给搭档沃克斯的信中写道，“景观（landscape）不合适，建筑（architecture）也不合适，两者结合起来仍然不合适——造园（gardening）更糟糕……这种艺术既不是造园也不是建筑。特别是对于我正在加利福尼亚州做的这件事而言，两者都不是。它是森林的艺术，是一种纯艺术，有别于园艺、农业，或者叫森林实用艺术……如果你一定要开创这样的新艺术，就不会想给它起一个旧名字。”

尽管这个名字造成了种种问题，它还是得到了保留。景观设计师也一直受到各种误解的困扰。比如，人们认为景观设计学是建筑学的一个分支学科，而不是一个凭自身力量独立出来的学科，因此，景观设计师是一种专科建筑师，就如外科医生是专科医生一样。再比如，人们觉得景观设计师就是风景造园师

图1　纽约中央公园鸟瞰图，最初依照弗雷德里克·劳·奥姆斯特德和卡尔弗特·沃克斯1858年的竞赛胜出方案进行布局

（这是一个常见的错误）。许多景观设计师会告诉你，有些时候
周围的朋友会邀请他们“为花园提一些建议”。我的一位前同
事曾回答说，“好啊，我想看看你的花园，但我必须先完成对风电
iii 厂的视觉影响评估”，然后享受对方困惑的眼神。虽然景观设
计师有时**确实**会设计花园，但这只是他们工作内容的一小部分。
因为景观设计师的工作包括了商业园区的布局规划、工业废弃
地的复垦、城市历史公园的修复，以及重要基础设施（如高速公
路、水坝、发电站和防洪设施等）的选址和设计。这本通识读本
的首要任务就是回答这样一个问题：景观设计学是什么？当代
从业者的工作范围相当广泛，就连奥姆斯特德“森林实用艺术”
的观点也无法涵盖。

这个问题有多种回答，我将把它们糅合起来。首先是从历史的角度来看，不仅着眼于景观设计学的根源，也关注奥姆斯特德与沃克斯所开创的这门年轻学科的成长、发展和传播。另一个角度是考虑景观设计师在当代社会扮演的角色，他们所承担的委托类型，以及与建筑师、城市设计师、城乡规划师、环境艺术家之类的其他专业人士之间的关系。第三个角度，也是我最感兴趣的角度，则是调查该学科的理论基础，还有各种美学、社会和环境论述。这些论述塑造了景观设计学这个学科，并把它从同类的领域中区别出来。于是，“它是什么？”这个问题逐渐演变为“为什么要这样做？”和“为什么它如此重要？”。

作为一个在英国学习的景观设计师，我难免会从英语世界的角度来撰写本书（英美两国的专业有共同的根源，而澳大利亚、加拿大和新西兰等英联邦国家之间也有着密切的关联）。然而，我也会通过在其他国家发现的不同起源和视角来试图让这

一点不那么明显。法国、德国、荷兰和斯堪的纳维亚都在景观设计学的发展中发挥了重要作用，但目前，景观设计学发展最快的 iv
地方似乎是中国。中华文明中的造园传统可以说与西方一样悠久，但中国人直到近些年才接受景观设计学，这与中国自1979年来的经济腾飞所带来的物质和社会巨变息息相关。有意思的是，我们注意到西方的景观设计学理念正在中国文化中发挥作用。不久，我们也许就能看到中国式思维和实践对世界其他地方的景观设计学实践方式所产生的影响。

与景观设计学相关的术语还是一片混乱，在进行跨文化的描述时更是如此。即使是在英语国家，术语的使用也各不相同，而在尝试用其他语种，例如德语或法语构建相关术语时，问题则更加棘手。仅仅为了解释该问题，我就能写完这本通识读本的三万五千字。虽然我不能避免对定义和意义上细微差别的些许讨论，但是如果用几个关键词就能阐明我的意思，那还是很有用的：

景观。这是一个模棱两可的术语，但《欧洲景观公约》中有一个被广泛认同的实用定义，它规定，景观是“一个为人所感知的区域，其特征是由自然和（或）人类因素作用与相互作用的结果”。这个定义的价值在于它同时包含了两个概念，即景观是一片土地，换言之，是某种物质，但它也是某种“为人所感知”的东西，能够由心灵和社会共享。

景观设计学。世界景观设计师与风景园林联合会是这样描述的：“景观设计师就户外环境、空间（建成环境内外）的规划、设计和管理，及其保护和可持续发展进行研究并提供建议。景观设计师这一职业需要拥有景观设计学的学位。” v

景观设计。由于“景观设计学”这个词并不完美，一些人更愿意选用“景观设计”这个术语。这是一组近义词，但景观设计可能会将“景观规划”（见下一条目）一词的内涵排除在外。相比之下，“景观设计学”这一术语的含义更广泛，也是国际劳工组织认可的行业名称。在美国，这两个词在法律上存在区别。景观设计学是一个由国家监管、实行注册制的行业，从事这一行业需要接受特定教育并顺利完成注册考试。景观设计则不受国家监管，也不需要特定的专业资格。

景观规划。这个实用的定义是由联合国教育规划署提出的：“在土地利用规划过程中涉及的物质、生物、审美、文化和历史价值，以及这些价值、土地利用和环境之间的关系与规划。”

总而言之，“景观设计学”是一个总的学科和行业名称。设计和规划之间存在交集，二者都是景观设计学的一个方面。

第一章

起　源

“景观设计学”一词最早用在出版物上，大约是在1828年吉尔伯特·莱恩·梅森的《意大利著名画家笔下的景观建筑艺术》一书的标题中。梅森是一位绅士型学者，他人缘很好，苏格兰最畅销的小说家沃尔特·斯科特爵士就是他的朋友，不过，梅森自己却没有什么著名的追随者。他用landscape architecture一词指代景观中的一系列建筑，而非景观本身。如果不是一位名叫约翰·克劳迪乌斯·劳登（1783—1843）的苏格兰同胞采用了这一表述的话，我们或许就再也听不到这个词了。劳登是位多产的设计师、作家兼编辑，1826年，他创办了颇具影响力的《造园师杂志》。许多人都读过劳登的书，包括他的美国同行安德鲁·杰克逊·唐宁。唐宁所著的《论风景造园的理论与实践》共发行了四版，售出约9 000册，其中一段名为“景观设计学或乡村建筑”。“景观设计学”一词似乎就是由此传入美国，随后又被弗雷德里克·劳·奥姆斯特德和卡尔弗特·沃克斯采用。

不过，要是“景观设计学”这一表述直到1828年才被创造

出来的话，我是怎么在前言中断定这门学科与建筑学和工程学
1 一样古老呢？在1975年首次出版的《人类的景观》中，作者杰弗里·杰里科与苏珊·杰里科深入调查经过设计的景观的历史，并用插图展示了布列塔尼地区卡纳克镇里上千块纪念碑和巨石的排布，以及威尔特郡巨石阵里50吨重巨石的布局。这些调查表明，自史前时代以来，人类一直在有意识地修整土地。同样，园林史的相关书籍中往往也在开篇就提出设想：最初的人类通过在土地周围设立防护屏障，创造了最早的院落和花园。正如我们所见，景观设计学通常涉及功能性和生产性景观的设计，如农场、森林和水库，但它与造园在美学、愉悦感和舒适性方面有着共同点，这不仅让它同最早的定居点和耕地联系了起来，也使它与古人对天堂乐园的梦想有了关联。

天堂乐园的构成总是取决于当时的条件。对于要在尘土飞扬、没有河流的高原上忍受严酷环境的古代波斯人而言，水显然是生命的源泉。他们发明了名为“坎儿井”的地下水渠，以此来补充灌溉渠，并把花园集中建设在交叉的水渠上，开创了经典的四分制设计——查赫巴格。花园被围墙封闭起来并与外部的沙漠隔绝，里面满是沙漠居民喜爱的元素，如椰枣、石榴、樱桃、杏等可作水果和提供阴凉的树木，还有凉亭、芳香灌木、玫瑰和各种草本植物，以及水池和喷泉等。我们现在所用的“乐园”（paradise）一词可以溯源到古伊朗语言（阿维斯陀语）中描述这种特殊花园的词pairi-daeza，该词后来被缩短成了paridiz。把称作“第一自然”的野外与人类定居和耕作的“第二自然”区分开来是很有意义的一件事，园林史学家约翰·狄克逊·亨特曾建议用“第三自然”一词描述公园和花园等带有特定美学意图

的地方。我们可以看出，景观设计学不仅涉及第二自然，也事关第三自然。至于这里面是否有可以被称作“第一自然”的部分，目前仍存在着争议。一些地质学家认识到了人类对大气层和岩 2
石圈的影响程度，已经将我们现今的时代称为“人类世”。人类对地球影响深远，足以让所有人类都在不安中意识到我们共同的责任，但这也印证了杰弗里·杰里科在《人类的景观》中的观点，有朝一日，“景观设计可能会被公认为最全面的艺术”。

直线与曲线：规整与非规整

在这里，我们首先需要概述一下园林史，因为景观设计师继承了几个世纪以来由造园师所承担的空间调查和实验，当他们为新的设计挑战寻求解决方案时，常常会借鉴或反对这些长期以来的传统。根据风格，花园可能有不同的分类，但从设计的角度，将它们视作一个连续体会很有帮助。这个连续体的一端是规则式花园，其特点是几何图形、直线和规则的平面布局，而另一端是非规则式或自然式花园，特点是形状不规则、曲线和更加丰富多样的平面布局。在这两极之间则存在着无数的变异和混合。比如在爱德华七世时期，英国工艺美术风格虽然以直线、规则几何和规整的平面布局为特色，却在种植方面体现出自然主义的柔和，同时还在所有铺装、墙壁或其他建筑元素中采取了利用当地材料和传统施工技术的乡土细节设计。

历史上最早的花园大多是规则式的，显然，用直线和直线形的模具进行测量和放样要简单得多。米利都（今土耳其境内）和埃及的亚历山大之类的古城都建立在网格状规划图上，几百年后，许多美国城市也采用了同样的规划形式。用规则的砖石 3

更容易建造建筑，两地之间直线的路径最短，平直的犁沟比弯曲的更容易挖掘，排水更有效，水渠和排水沟也同理。尽管人类在散步时习惯走略微弯曲的弧线，但是仪式中的游行却更有可能沿着直线行走。景观史学家诺曼·牛顿认为，轴线这种最能体现空间秩序的元素起源于穿过寺庙场地的游行路线。在规则式设计中，主轴是一条想象中的线，垂直将建筑的正面一分为二。这些建筑也许是寺庙、教堂，也许是一栋大房子。轴线连接了两个点，为两侧对称提供了可能，即图纸的一侧与另一侧镜面对称。在整个文艺复兴时期，欧洲花园都有这个特点，如意大利巴涅亚附近的兰特庄园和巴黎的卢森堡花园。在17世纪法国路易十四时代的造园大师安德烈·勒诺特的作品中，这种组织花园空间的方法体现得最为典型。在离巴黎12英里外的凡尔赛宫，勒诺特设计了占地面积约为纽约中央公园两倍的花园。对植物的处理同样沿用了这种规则式的设计，这些植物被反复修剪，直到它们成为绿色的砖石。在巨大的人力和物力消耗下，自然受到了严格的操控，哪怕是路易十四本人都无法随心所欲：不管他雇用多少工程师，派遣多少士兵去建设沟壑水渠，都无法让他的喷泉全天候运转。凡尔赛宫成了全欧洲许多皇家花园的典范，维也纳的美泉宫、圣彼得堡郊外的彼得大帝夏宫以及伦敦附近的汉普顿宫等著名皇家花园都以此为范本。

到了18世纪的英国，花园设计师和他们的赞助者不再拘泥于法式的规整和拘谨，转而青睐那些随着近百年的发展越来越趋于不规则和自然式的设计。对于这种变化，人们有着不同的
4 解释，一方面是来自荷兰设计的影响，另一方面是对于中国传统的报道。英国的地主当然希望远离法国的严谨和规整，他们将

这些与令人憎恶的君主专制联系在一起。英国的赞助者往往崇拜画家克劳德·洛兰和尼古拉斯·普桑所绘的风景画，这两位画家都在罗马度过了大半生，喜欢以罗马坎帕尼亚平原的风景为灵感，创造出世外桃源般的场景。更抽象地说，非规则式在花园设计中的兴起与人们对经验主义日益浓厚的兴趣不谋而合。对有理几何的执着让位于对自然界表面不规则性的仔细观察。威廉·贺加斯在《美的分析》中确立的蛇形“蜿蜒线”与兰斯洛特·“万能”布朗所设计的湖边曲线非常相似。到了18世纪中叶，布朗（1716—1783）的地位不断提高，直到现在仍是其同行中最令人印象深刻的人，一方面是因为他作品丰富，另一方面则是由于这个难忘的绰号。布朗的绰号源于他的一个习惯：在参观完赞助人的庄园后，布朗总是告诉他们，他从中看到了“能力”，用他自己的话来描述，叫作“可能性”或“潜力”。布朗的设计模式包括去除露台、栏杆和所有规则式的痕迹，在公园周围种植林带，给河流筑坝形成曲折的湖泊，以及将漂亮的树木以孤植或丛植的方式点缀到公园中。有意思的是，布朗并不把自己称作风景造园师，他更喜欢“场地创造者”和“改良者”的称呼，这两个词语在很多方面比“风景造园师”更接近现代景观设计师的角色。在威尔特郡的朗利特庄园、西萨塞克斯郡的佩特沃斯庄园和利兹城外的坦普尔纽萨姆庄园都可以找到典型的布朗风格。

对布朗的批评始于他所在的时代，并在他去世后愈演愈烈。他在当时受到批评，不是因为破坏了许多规则的花园（虽然确实如此），而是因为他对自然的了解还不够深入。在他的批评者中有两位赫里福德郡的乡绅，也是新的如画风格的倡导者——

5 尤维达尔·普赖斯和理查德·佩恩·奈特。一处风景或设计如果想成为如画风格，必须是适合风景画的选题，但热衷新时尚的这些人认为布朗的风景太枯燥了，并不符合这一要求。奈特的说理诗《风景》就是针对布朗而作，他曾说布朗的干预只能创造出一片“沉闷、无趣、平淡而静止的景色”，需要增添一些草木丛生且多样化的粗犷风景才好。这一争论如今也反映在了修剪后的草坪与野花草地之间的对立中。在美国，修剪整齐的草坪一直是人们对于前院的正统处理方式，通常受到城市法令的管制。除了受到精心照料的单一草坪，在房前种植其他任何植物都可能引发争议。

布朗自封的接班人汉弗莱·雷普顿（1752—1818）与如画风格的热衷者们争论不休，但公众却在很大程度上受到了校长兼艺术家威廉·吉尔平（1724—1804）的影响。吉尔平出版了一系列瓦伊河谷和英国湖区的游记，使公众对如画风景产生了无法抑制的兴趣，这种审美至今仍占据着主导地位。然而“如画”一词如今已经失去了原意，很少再以大写字母的形式出现。对许多人而言，它现在只意味着“漂亮”或“迷人”，与绘画之间已经失去了联系。

然而，雷普顿在景观设计学的开创中有着特殊的地位。他是第一位称自己为风景造园师的从业者。在他决定效仿布朗之前，他曾尝试过许多职业——记者、剧作家、艺术家和政治代理人。他并没有什么深厚的园艺知识，但他想出了一种巧妙的方式，将自己的想法按照改造前和改造后分别绘制成水彩草图，装订在红色封面中呈现给客户（图2a、2b）。通过翻动折页，客户可以清楚地看出雷普顿对他们的庄园提出的改造建议。这些

图2a “改造前”的全景图，摘自汉弗莱·雷普顿为安东尼别墅所作的《红皮书》，约1812年

图2b “改造后”的全景图，摘自汉弗莱·雷普顿为安东尼别墅所作的《红皮书》，约1812年

别出心裁的《红皮书》是当今可视化方法的先驱——比起水彩 6
画，如今更常用的是计算机模型和漫游动画。布朗和如画风格的提倡者分别用自己的方法把想法灌输给各自的客户。雷普顿更像是现代的景观设计师，他了解客户的需求，听取他们的

7 意见。因此，他和布朗的模式分道扬镳，重新在房子附近布置露台，形成了一种实用的花园特色。他写道："我发现在人类居住区附近，实用性常常比美学优先，而便利性比如画效果更受人青睐。"

"风景造园"转变为"景观设计学"

无论是劳登还是唐宁，两人的著述标题都使用了"风景造园"（landscape gardening）一词。在英美传统观念中，风景造园被认为是景观设计学的先驱。前者服务于私人客户，而后者通常提供公共服务。以伦敦东区和英国北部工业城市为代表的公园建设运动促进了这一变化。作为社会改革运动的一部分，公园建设运动始于19世纪30年代，与哲学家杰里米·边沁的功利主义精神一致。劳登是这位哲学家的友人，同时也是埃德温·查德威克和议员罗伯特·斯莱尼的朋友，前者曾为公共卫生改革而奔走，后者则在议会为公园做过辩护。基于最多数人的最大幸福感的功利主义观点，仍然是建筑环境中许多设计和政策决定的基础。19世纪的立法为英国地方政府建设市政公园创造了条件，这些公园很快就成了市民的骄傲。英国中部的德比植物园便是其中最早的公园之一，而它就是劳登的设计作品。这座植物园是一位慈善纺织品制造商兼前市长赠送给这座城市的礼物，顾名思义，它以收集树木和灌木为特色，并贴上了教育用途的标签。公园可以改善人的身体和精神状况，而在那个真的担心会爆发革命的时代，人们觉得不同阶级在公共场所的交流可以提升公共秩序。劳登放弃了对如画风格的追求，转而采
8 用一种由人工精心设计的方法来布局和种植，他称之为"花园

式景观学派”。它以几何种植床为特色，先在温室中培养外来植物，再移植到外界环境中。花园式风格很快就被维多利亚时代的公园所认可，为卓越的园艺水平提供了丰富的展示机会。约瑟夫·帕克斯顿（1803—1865）是与劳登同时代的杰出人物，他从一名在伦敦百灵顿伯爵大屋皇家园艺学会花园工作的卑微造园师成长为创造1851年世界博览会水晶宫的著名设计师。他承担了许多公众公园的设计，但其中最关键的作品是坐落于默西赛德郡的伯肯海德公园。1850年，奥姆斯特德在访问英国时参观了该公园，从中获得了设计中央公园的灵感。如果恰和时宜的话，将劳登和帕克斯顿之类的设计师称作“景观设计师”并没有什么问题，但在他们的时代，这个词还没有被创造出来。

虽然现如今把自己称作“花园设计师”可能更流行，薪水也更丰厚，然而与此同时，“风景造园师”的头衔并没有消失。就像我接下来将要展示的那样，尽管花园设计师可能像大厨一样更为公众所熟知，但景观设计学是一个更广泛的领域。如今，花园与景观设计师的关系就如同私人住宅与建筑师的关系。建筑师有时会设计私人住宅，景观设计师有时也会设计私人花园并在切尔西花展上展出。不过，他们的主要生计还是依赖更宏大的项目，并且无论如何都与开发有所关联，这在大型机构中尤其明显。直到1899年，美国景观设计师协会在纽约的一次会议上成立时，景观设计师才被正式确立为一种职业。有趣的是，他们将承包商、建设者和苗圃工人排除在外，却把比阿特丽克斯·法兰德纳入景观设计师的范畴，而这名设计了华盛顿特区敦巴顿橡树园著名花园的设计师，直到她杰出的职业生涯结束时，还一直坚持称自己为“风景造园师”。在英国，景观设计师协会（现称景观协会）直到1929年才成立，这时离奥姆斯特德和沃克斯在竞赛中引入这一称谓已经过去了71年。

在其他地区的起源

对景观设计学的基本描述无疑是一个跨越大西洋的故事，而奥姆斯特德的伯肯海德之旅常常被誉为这一故事的开端。但是在其他国家，我们也可以找到相似的历史。我将列举一些出现在欧洲的例子，来说明景观设计学如何从早期的花园和公园设计传统中脱离出来，它们也将说明不同的历史文化特色如何塑造了这一学科在每个国家的发展特点。在18、19世纪的法国，英式花园风格被广泛采用，工程师让-查尔斯·阿道夫·阿尔方（1871—1891）在园艺师让-皮埃尔·巴里耶-德尚的支持下，结合奥斯曼男爵的巴黎改造，建造了一系列与之相关的公园。其中最引人注目的是位于城市东北部的肖蒙山丘公园，里面有座曾经的石灰岩矿场。在矿场高耸的峭壁顶上，有一座罗马灶神庙的复制品。“景观设计师”在法语中叫作paysagiste（最贴切的翻译是“乡村主义者”），但一直到第二次世界大战后，凡尔赛的园艺学校开设了第一批培训课程，该职业才被官方认可。在20世纪后期的几十年中，重新兴起的法国传统规则式花园通过与混乱的后现代思想糅合而获得了新生，并由于为枯燥的如画风景创作提供了大胆的选择而风靡一时。

在德国，风景造园向景观设计学转变过程中最重要的人物是普鲁士国王的一名园丁，彼得·约瑟夫·莱内（1789—1866）。除了皇室的委托之外，他还设计了德国最早的一批公园，包括马格德堡的腓特烈-威廉公园、法兰克福的奥得河畔莱内公园以及柏林的蒂尔加滕公园和弗里德里希斯海因人民公园。直到1913年，德国花园建筑师协会才在法兰克福成立，又在1972年更名为德国景观设计师协会。在魏玛共和国时期，开放式公共绿地的设计受到关注，但是景观设计学的发展受到了活跃的纳粹主义实践者的影响。在这一时期，景观设计师的工作不仅是沿着新建的高速公路植树，更臭名昭著的是，他们会在被征服的东部地区开展乡村景观的“德国化”。第二次世界大战后，景观设计学起码在西部地区迅速重组，从业者们在重建被战争破坏的国家中扮演了重要的角色。自联邦园艺博览会1951年在汉诺威首次举办后，这个两年一度的展会持续展示了景观设计学如何将废弃地和战毁地变成永久的公园。

荷兰的公园设计者小扬·戴维·措赫尔（1791—1870）受到了布朗和雷普顿的影响，他设计的阿姆斯特丹冯德尔公园于1865年开放，是一座英式的浪漫自然主义公园。然而，荷兰也有填海造陆的历史，因此，到了20世纪，需要进行全面的景观规划，从而在圩田上创造全新的居住区和景观。被誉为荷兰生态运动之父的植物学家雅各布斯·彼得·蒂塞斯（1865—1945）提出了一个具有国际影响力的观点：他建议每个城镇或社区都应该有座“教育花园”，使人们在家门口就可以了解自然。蒂塞斯很担忧排干沼泽和荒地造林等人类行为造成的乡村物种减少。荷兰已经高度城市化，显然，乡村也是人类创造的产物，但人们却渴望与自然接触。也许正因为如此，这个国家在景观设计学和都市主义方面诞生了一些最为有趣的新构想。在这里，“景观设计师”这个头衔自1987年就开始受法律保护了。

如今，景观设计学已经成为一门全球性的学科，然而在许多

11

国家，它依旧处于起步阶段。超过70个国家级协会都隶属于世界景观设计师与风景园林联合会，名单依字母顺序，从阿根廷和澳大利亚开始，排到乌拉圭和委内瑞拉结束，其中包括了美国、中国和印度这样的人口大国，也包含了拉脱维亚和卢森堡这样的小型国家。不同国家中，从业人员的数量也千差万别。加拿大景观设计师协会拥有超过1 800名成员，法国景观设计联盟拥有超过500名成员（但只代表了其中三分之一的从业者），德国景观设计师协会有大约800名成员，英国景观协会有超过6 200名成员，而1992年刚成立的爱尔兰景观学会仅有160人。目前，最大的协会是拥有约15 500名会员的美国景观设计师协会。虽然各协会在教育、资格认证和注册要求方面努力实现标准化，但由于地方体制结构和法律的不同，各国间的要求仍然存在着很大差异。即使在教育方面，也存在诸多不同。在一些国家，景观设计学授课与园艺学、农学和造园学相结合，在另一些国家则与建筑、规划和城市设计相配套，而在其他一些地方，它可能会出现在与林学或环境科学相关的学校中。虽然在这些不同类型的机构中，景观设计学课程方案的相似性远远超过了差异性，但毫
12 无疑问的是，它们各自都有着不同的侧重点或特色。

在读完这段简史之后，我希望你能够对景观设计学的范围有一定的了解，但是，此刻你也许想知道这门学科是否有什么明确的核心。下一章我们将会讨论，在景观设计学的总体框架下，各种人类活动的具体范围，还将关注对该学科的本质进行定义
13 的各种尝试。

第二章

景观设计学的范围

现在，是时候来看一下景观设计师到底在做什么了。我们研究这一小部分案例，旨在传达当代景观设计学实践的多样性，希望能让你对该学科的大致范围有所了解。本章介绍了四个项目，包括备受瞩目的总体规划、视觉影响评估、艺术化的城市设计，以及社区参与活动。从重视实用价值到充满天马行空的想象，这些项目各有不同。你或许可以思考一下，它们是否都属于“改良和场地营造”的范畴。

新加坡滨海湾花园（2006年至今）

第一个项目是新加坡国家公园局组织的一场设计竞标的产物。当时，他们希望通过这一项目来寻找设计团队对滨海湾花园做总体规划。作为一个以园艺为主题的景点，滨海湾项目坐落在滨海湾新城区，是一片因填海造陆而形成的滨水地带。这里最终将建设超过100公顷的热带花园，包括滨海湾南花园、滨海湾东花园和滨海湾中央花园。滨海湾南花园是这个大型项目

的第一阶段，被委托给了由格兰特景观事务所和威尔金森·艾尔建筑事务所联合组成的英国设计团队。格兰特景观事务所的
14 总体规划灵感来源于新加坡的国花——兰花的形状。该方案得到了最高行政级别的支持，它将自然和技术相融合，包括了由建筑师设计的“花穹”和“云雾林”两个人工生物群落，分别容纳生长在地中海地区和热带山地气候区的植物。景观设计师设计了引人注目的“擎天树”，其中有些高达50米，它们既是植物冷室中冷却系统的一部分，同时也高高地生长着一些用于展示的附生植物、蕨类和开花的攀援植物，在夜晚还能发挥照明作用。隐藏在“擎天树”中的科技成果模拟了真实树木的生态功能，比如它们装有光伏电池，可以为部分照明设备供电；能够收集和引导雨水，以用于灌溉和水景表演等。

人们将滨海湾花园描绘成园艺界的迪士尼乐园，拥有《爱丽丝梦游仙境》中的场景。同时，它也被认为是环保设计的胜利。不出意料，它引起了全球媒体的注意。即使对于最著名的设计事务所来说，具有如此规模和雄心的项目也十分罕见，但滨海湾花园仍然体现了当代景观设计学实践的许多特色。比如该项目拥有一个涉及多学科的设计师团队，其中不仅有景观设计师和建筑师，也包括专业的环境设计顾问、结构工程师、游客中心设计师和通信专家。场地的位置和条件——填海造陆形成的滨水土地——也是近几十年来许多大型项目的特色，丹麦哥本哈根的海港公园（2000年竣工）、瑞典马尔默丹妮娅滨海公园（2001年竣工）以及位于中国上海的后滩湿地公园（2010年竣工）都与此类似。

图3　新加坡滨海湾花园项目的第一阶段委托给了由格兰特景观事务所和威尔金森·艾尔建筑事务所联合组成的英国设计团队 15

威尔士赫迪威尔风电厂景观与视觉影响评估（2010）

虽然滨海湾花园引人注目，成了一项标志性设计，但对下
16 面这个属于战略性景观规划而非总体规划或场地设计的项目来说，如果能让公众感觉不到它的存在，那就成功了。景观设计师的工作常常与对乡村干扰的最小化有关，比如他们经常参与矿场扩建，或是露天煤炭的相关申请等活动。近年来，风力机的选址已然成为英国土地利用规划中争论最激烈的问题之一。无论特定涡轮的设计有什么优缺点，在某些乡村地区，这些机器仿佛引发了一种几乎是发自内心的厌恶，这大概是因为人们将其看作外人强加给他们的负担，只对遥远的城市有利。尽管人们常常赞同清洁能源或高速运输之类的想法，但如果这是在自家门口提出的，他们就会站出来反对，这种现象被称作“邻避主义”［NIMBYism，即not-in-my-backyard（别在我家后院）］。提案中涡轮机的高度和数量会影响其被接受的程度，现有的地形也是如此。在风力机方案中，景观设计师已经成为为“视觉入侵区域”建模和制图的专家。事实上，对于景观中的任何大型添加物，景观设计师都会这么做。利用图像合成和计算机可视化技术，他们能够从多种关键视角展示所有提案的外观。他们还经常参与制订缓冲方案，其中可能包括筛查土方和种植工程等，以减少此类活动的影响。

从技术角度而言，开阔的高地特别适合风电厂选址，但这类场地往往因其现有的景观特征而受到高度重视。威尔士政府承诺，到2025年，可再生能源的发电量将增加一倍，并确立了七个可以开发大规模风电厂的战略研究区域。英国AMEC环境基

础设施有限公司是一家提供景观设计等一系列服务的公司，受NUON可再生能源公司委托，对其在波伊斯修建的赫迪威尔风电厂提案进行景观与视觉影响评估。这项评估很复杂，因为该公司在这一项目的东侧区域也规划了另一座风电厂，而且还在附近运营着另一处设施，他们希望用数量更少但更高大的涡轮机重新配置。顾问团队的可视化技术让这些方案的累积效应得到了评估，并使涡轮机的规划数量从13个减少到了9个。评估在2010年完成，但在撰写本文的时候，该方案仍在规划当中。

大型基础设施项目可以在全国范围内带来巨大收益，但也可能对当地产生重大影响，赫迪威尔风电厂就是一个很好的例子。在规划体系完善的国家，获得此类开发的批准可能是一个漫长而复杂的过程。景观设计师可以在各个阶段提供协助，从前期评估，到可能需要向公众质询提供证据的规划过程本身，再到缓解方案的设计和实施。值得一提的是，尽管私人开发商可以聘请景观设计师，但在许多国家，地方当局也可以聘请他们。就像很多法庭剧中，控辩双方将法医病理学家当作对手一样，景观设计师有时也会发现，自己在特别有争议的规划质询中处于对立的一面。

加拿大蒙特利尔玫红球（2011）

加拿大的景观设计师克劳德·科米尔（1960—　）以其对城市生活的诙谐和艺术化干预而闻名。其中的许多作品以成为城市肌理中的永久设施为目标而创作，如多伦多的糖果海滩和蒙特利尔旧港口钟楼码头街边的钟楼海滩，但他的临时装置作品也广为人知，玫红球就是其中之一。在2011年的自由节期间，

18 17万个粉色树脂球贯穿了位于蒙特利尔同志村的圣凯瑟琳东街，将这条平凡的街道变成了令人着迷的步行街。这些球共包括三种不同的大小，采用了五种略微不同的粉色。它们横跨整个街道，与林荫树的枝丫相互交织形成一片穹顶，在贝里街和帕皮诺街之间投下了绵延1.2千米的斑驳阴影。装置分为九个部分，沿途通过各式各样的图案营造出不同的氛围。

科米尔是美国景观设计师玛莎·施瓦茨（1950— ）的门生。玛莎·施瓦茨曾是打破学科传统的“叛逆型天才”，但现如今已经成为最受尊敬的教育者和从业者之一。入行前，施瓦茨拥有艺术学背景，她早期的作品融入了非传统的材料，比如塑料材质的树和花、有机玻璃碎片，甚至是令人诟病的涂漆甜甜圈，这些时常出现的蓄意挑衅招致了景观设计学界的拒绝：“这些真的算景观设计作品吗，还是些别的什么？”像施瓦茨、科米尔和德国Topotek 1事务所这样的从业者是景观设计学界中的顽皮派，他们喜欢推翻假设、打破预期。然而除了乐趣以外，其中还有对场地、内涵和用户需求的理解，当设计将会造成长期影响的时候更是如此。最好的设计实践认识到了这些所有的因素，因此这些作品不仅有趣，同时也具有实用性。

西费城景观项目（始于1987年）

西费城景观项目是一项由景观设计师、教育家、作家、摄影师兼活动家安妮·惠斯顿·斯本发起的行动研究项目。该研究最初设在宾夕法尼亚大学景观设计学和区域规划系，斯本自
19 1986年起在那里担任教授，直到2000年她转到波士顿的麻省理工学院才离职。该项目从一开始就追求将研究同教学和社区服

务相结合，还特别关注了对西费城贫困社区中一系列社区花园的设计和施工。这些都是渐进式的小规模改善，无法解决住房存量不足、基础设施缺乏、贫困和失业等所有问题，但除了为城市景观增色，它们还催化了其他形式的社区发展。1995年之后，该项目衍生出一个分支，叫作米尔溪项目，由宾夕法尼亚大学的学生和研究人员与西费城苏兹伯格中学的教师和学生合作完成。它围绕一门名叫“城市流域”的中学新课程的开设而组织，旨在提高学校所在地区的环境意识。项目以一条地下涵流——米尔溪为中心，如今学校的操场就建在从前米尔溪曾经穿过的田野上。这条被埋在地下的河道曾经引发洪水、沉降和彻底坍塌等诸多问题。参与项目的景观设计师可以提醒人们，使他们注意到在洪泛平原上进行城市开发会带来什么样的麻烦。他们还提出了重新设计未开发土地以滞留雨水的方法，从而在提供有社会价值的开放空间的同时降低洪水风险。

我无法给出西费城景观项目的结束时间，官网上的时间线在2009年就结束了，但它们的博客仍会偶尔发布帖子。看起来，该项目对这些问题社区和居民的影响可能会持续几代人，而建设的花园则会成为一笔有形的遗产。不管用什么标准衡量，西费城景观项目都算得上是景观设计学界持续时间最长的社区参 20
与项目之一，当然，也是最受认可和称赞的一个。2001年在白宫举行的由40位主要学者和艺术家参加的公共生活领域峰会中，该项目获得了“最佳实践典范”称号。2004年，这一项目又获得了由美国景观设计师协会颁发的社区服务奖。斯本最近的著作《自上而下还是自下而上：重建社区景观》就是以她从事该项目25年的经验为基础撰写的。

它们有什么核心吗？

在本章中，我专门选择了四个项目展开讨论。在这些项目中，除了景观设计师都发挥了主导作用以外，它们看起来并没有什么其他相似之处。但通过进一步思考，也许我们就可以发现共同点。滨海湾花园和玫红球的设计者都试图创造出能吸引游客的视觉奇观和节日感。西费城景观项目和威尔士风电厂的研究则都关注开发选址和基础设施建设的后果。然而，我们很容易就可以找到另一批像这四个项目一样彼此不同的景观设计项目，然后再找到另外的四个……项目的多样性反映在了不同的设计方法中。一些景观设计师以其作品的隐蔽性为傲，在减弱拟建设的高速公路或输电线路的影响时，他们希望自己的作品可以尽可能和谐地融入周围的景观中；另一些设计师则追求惊人、有趣或是戏剧性的效果，如果发觉自己的艺术性被忽视了，他们就会感到很沮丧。有些设计师非常重视与社区合作，以此升华任何倾向于支持社会可持续成果的自我主义冲动；另一些设计师不能忍受妥协，他们认为最好的设计作品表达了一个独特的视角；还有些设计师可能会结合实践的特点，或者根据场
21 地、客户或设计概要的不同来改变方法。

既然存在这么多不同的观点，且景观设计师从事的项目种类又如此多样，那还有没有可能对这个学科提出明确定义，或是说明它的本质吗？在我看来，对于定义和界限的要求通常是一种不安全感的表现。对一门学科的专业化涉及通过考试来制定入行标准，因此亟须确定从业者应具备的核心知识和技能。这样做有利于为客户和公众提供保护，专业人士应该知道自己在

做什么。在医学领域，这种情况很容易举例，没有人会信任一位没有执照的脑外科医生；在土木工程领域，没有进行必要的计算就建设桥梁显然也会影响公共安全，但是如果有大量可以在社会中学习相关技能的机会，且一些潜在的危害比较分散、短时间内不好发现的话，那就不太容易界定了。建筑师、城市规划师和景观设计师在不同程度上参与公众质询和参与式交流的事实表明，非专业的知识和意见也会得到重视。甚至可以说，该学科就是建立在专业化的非专业知识基础上。的确，在规划或设计不当的住宅区及新城镇中，一些长期危害常常被归咎于没有理解人们真正想要或需要从这些发展中得到什么。专业化的缺点在于可能导致保护主义态度，或是排他性的“封闭型工厂”，将客户、使用者和类似角色的人员拒之门外。许多关于标准、守则和认证的讨论都是为了限制特定的工作领域，同时由于受到商业利益驱动而常常令人怀疑。

专业化的一个体现是尽快确立核心课程体系，但对于景观设计学这样多元又内涵广泛的学科而言，这已成为一个不可能
的要求。景观设计学也许有一个流动的核心，但没有固定的本 22
质。它与包括工程、艺术、建筑、城市规划和城市设计在内的其他学科之间存在边界，但这边界不是固定的，而是互相渗透。然而，它仍然是一门独立的学科，不能被相关学科吞并。一个将景观设计学概念化的有效办法是将它看成一个大家族。在这个家族中，有些人做着与弗雷德里克·劳·奥姆斯特德和卡尔弗特·沃克斯相同的事情：他们负责设计公园或者公园系统，尽管他们不一定认同奥姆斯特德关于城市中心田园风光适宜性的观点。还有一些人从来没有设计过公园，即使他们曾与设计交通

基础设施的工程师共事。其他人则专注于私人客户，几乎完全从事花园的设计工作。有些人的职业生涯与林业工作者一起度过，帮助他们以视觉与环境和谐的方式来开展种植及经营。而对于另一些人，当他们在城市工作、参与城市广场和步行街的设计或翻新时最快乐。从这些范围中选择两个人，你会发现他们的工作区别很大，甚至很难想象他们从事的是同一个职业，但是相似的网络把他们彼此联系了起来。景观设计学的开放性也许是它最大的优势，而它渗透性的边界也应当成为其他学科的典范。

为定义该学科所做的尝试通常都会失败（包括我在前言的结尾中引用的世界景观设计师与风景园林联合会所做的定义），而且我觉得这些失败不可避免。这些定义大多冗长而繁杂，试图把景观设计师参与的所有活动都囊括进去。已故的哥本哈根大学景观设计学教授马莱内·赫克斯娜撰写过一本书《向天空敞开》，据说书中写道，景观设计学关注所有没有屋顶之处的规
23 划和设计，但即使是这样内涵丰富的定义也存在着失误，因为整本书写的都是**室内**景观设计。这就是本质主义定义的问题：对于这些定义，人们通常能找到不合适的反例。我很喜欢伦敦格林威治大学教授汤姆·特纳的观点：景观设计学就是“建造优质的场所”。他突出“优质”这个词是为了强调建造任何的老旧场地都不算——当然他的叮嘱很笼统，而且留下了什么才算是优质的问题。这个问题的答案可能与生态学、心理学、社会学、政治学、美学，以及其他学科都有关联。这些是在21世纪进行场
24 地营造和改善时需要列入考虑的因素。

第三章

现代主义

1899年，当景观设计师这一职业在纽约会议上确立时，对传统的排斥已经席卷了整个艺术世界。对于不同的学科，“现代主义”有不同的含义，但都会关注探寻同工业化和技术进步所带来的新社会条件相关的表现形式。现代主义思想在美术（特别是绘画）和建筑中的轨迹截然不同，但两者都对景观设计学这门新兴的学科产生了有力的影响。

现代艺术的影响

20世纪英国的元老级景观设计师杰弗里·杰里科爵士（1900—1996）提出，景观设计与视觉艺术，特别是绘画有着特殊的关系。他认为，一项景观设计需要很长的时间来创造，因为即便设计阶段可以快速完成，在施工阶段也可能涉及大量土方的移动、大型湖泊黏土衬层的搅拌捣实和上百株树木的种植等，这些往往很耗时，远远超出了任何个体的单独工作能力。即使地形已经完工，植物也种好了，但可能还需要经过好几个生长季

25 才能达到预期的景观效果。这些限制使实践变得困难。另一方面，画家却处于一个相对令人羡慕的境地，他们对材料的需求更少：一间工作室、一个画架、几张画布和几支颜料就够了。因此，杰里科认为画家可以充当美学的探路者，而景观设计师所能做的最好的事就是跟上他们的步伐。在18世纪，景观的设计者密切关注各种艺术作品，从尼古拉斯·普桑（1594—1665）、克洛德·洛兰（约1604—1682）和萨尔瓦多·罗萨（1615—1673）等著名画家的画作中汲取灵感。然而到了19世纪，事情开始变糟了。由于热爱冒险的植物猎人把大量新物种和新品种带回英国，再加上技术的进步，如蒸汽加热的玻璃温室的发展，造成了人们对园艺的过度热衷，风景园林和绘画的联系因此被切断了。不过到了20世纪，景观设计与艺术重新建立了联系——但与此同时，艺术已经向前发展了。

即使是在那个风景画仍旧是流行画派的时代，艺术界也普遍不重视对地形的准确描绘。正如地理学兼艺术史学家彼得·霍华德所见，如果想查找一幅精确记录的风景画作，相比知名艺术家，你更有可能在没那么知名的艺术家的作品中找到。亨利·富塞利并不重视逼真性，因此在他担任皇家美术学院秘书时，会排挤那些“平淡描绘了某个地点”的作品。无论如何，记录风景的任务交给了可以做得更准确也更快的摄影师。有着严肃目的的艺术家们对摄影的到来做出了回应，他们选择转向了抽象艺术。像霍华德所指出的一样，风景画有立体主义、超现实主义和表现主义等等，但是对于场所的描绘（如果可以使用这
26 个词的话）远远比不上将要探索的理论和方法重要。然而杰里科相信，正是抽象艺术能为景观的设计者指明前进的方向。

有些景观设计作品确实从抽象画中获得了直接的灵感。建筑师加布里埃尔·盖夫莱康（约1900—1970）在1925年为国际装饰艺术和现代工业博览会设计了《光与水的花园》，如今被认为是装饰艺术的样板之作。这件作品也打动了查尔斯·德诺瓦耶，使他把自己在耶尔地区别墅的三角形抽象花园委托给了盖夫莱康设计。装饰艺术受到了立体主义和技术狂热主义的影响，但它缺少对功能的关注，因此有别于其他的现代主义建筑思潮。盖夫莱康的三角花园使用了混凝土，并采用几何图案和稀疏的种植方式，这些做法一定会让许多园艺家感到困惑，但设计师将其视为对园艺和自然主义传统的绝对突破。然而这些花园受到赞美主要是因为其风格，而不是因为它们是可以使用的"室外房间"。尽管如此，美国景观设计师弗莱彻·斯蒂尔（1885—1971）对盖夫莱康的作品印象深刻，他写了一篇题为《花园设计的新先锋》的文章，但这一现代主义的号召起初并没有引起同行的重视。斯蒂尔开始在自己的设计实践中尝试现代理念，这些尝试都是基于意大利的规则式或英国的自然风景式风格。在设计位于马萨诸塞州斯托克布里奇的纳姆柯基庄园（1925—1938）时，他采用了文艺复兴时期的理念，在林地中建造了一系列台阶，并创作了一个简单又上镜的现代主义版本——配有优雅白色金属扶手的蓝色阶梯。

另一位受到抽象化趋势影响的是巴西的博学家罗伯特·布雷·马克斯（1909—1994），他是一名画家、雕塑家、珠宝设计师、戏剧布景设计师，同时也是植物学家、育种家和景观设计师。他认为自己主要是画家，其多彩的油画风格与阿尔普和米罗有相似之处。在他的种植设计中，也可以发现生物般的造形和鲜艳的色彩。就像他为蒙泰罗一家在彼得罗波利斯附近设计的著名庄园（1946）所展示的那样，他可以通过单一色块，用植物的枝叶来创作隐喻性的绘画。他的一些作品高度图案化，比如他为故乡里约热内卢科帕卡瓦纳海滩长达三英里的长廊所设计的人行道（1970）就体现了这一风格。除了这些引人注目的私人委托以外，他还参与了几个著名的公共项目，其中有为奥斯卡·尼迈耶在里约热内卢承担的教育卫生部大厦设计的屋顶花园（1937—1945），以及与建筑师卢西奥·科斯塔合作完成的巴西利亚规划（1956—1960）。

另一位重要的过渡人物是托马斯·丘奇（1902—1978），他曾在加州大学伯克利分校和哈佛大学学习，后来获得了奖学金，前往西班牙和意大利。在那里，他发现加利福尼亚州的气候与地中海地区非常相似，有利于户外生活（图4）。在大萧条期间，他在旧金山开设了一家小型事务所，并逐步开启了自己的事业：为富裕的中产阶级客户设计花园，而不是替奢靡的大富豪服务。弗莱彻·斯蒂尔对现代主义的拥护使得丘奇更轻易地摆脱了来

图4　托马斯·丘奇的柯卡姆花园（1848）平面图：作为室外居住空间的花园

自对称性的束缚。他设计的那座有着观景木台和不规则形态游泳池的休闲花园成为西海岸生活方式中显眼的组成部分，很快， 28
这种设计方式不可避免地以“加利福尼亚式风格”而闻名。他通过各类生活杂志上的文章来宣传这一风格，《日落》是其中最有影响力的刊物，主要针对从东部地区移居到加利福尼亚州的人。丘奇受到了抽象艺术的影响：像布雷·马克斯一样，或许他也借鉴了阿尔普的图形，并采用了一种让网格铺装或锯齿形木凳与钢琴般的曲线互相映衬的独特设计手法。虽然丘奇受到了立体主义和超现实主义的影响，但1937年他与现代主义建筑师阿尔瓦·阿尔托在芬兰的会面推动了他的风格发展成熟。考虑到气候因素，他的花园很少采用草坪，因为在西海岸的气候条件下，草坪必须要持续灌溉才行。相反，丘奇使用了铺装、砾石、沙子、红木平台和耐旱的地被植物（图5）。他设计了约2 000座花园，但其公认的杰作是加利福尼亚州索诺马县的唐纳花园（1954

图5　托马斯·丘奇在加利福尼亚州索诺马县设计的唐纳花园（1954年与劳伦斯·哈普林共同设计）成为西海岸生活方式的象征

年与劳伦斯·哈普林共同设计），在该花园中，许多元素被和谐完
29 美地结合在了一起。丘奇沿着场地内现有的橡树布置宽大的木质平台，这一理念同样备受赞誉；阿达林·肯特为花园设计了优美流畅的池中雕塑，成为太平洋海岸享乐主义生活方式的象征。

建筑理论的影响

一些有影响力的景观设计师也是合格的建筑师，还有些可能与建筑师有着密切的合作。长期以来，这两个学科联系紧密，因此建筑理论的发展也不可避免地会对景观设计学产生影响。据历史学家尼古拉斯·佩夫斯纳所言，现代主义建筑起源于新艺术运动对过去束缚的拒绝，以及英国工艺美术运动对设计卓越性与完整性的追求。当这些潮流与工业技术的巨大潜力，还有钢铁、玻璃等新材料结合在一起时，打破传统的道路也就开通了。现代主义建筑比它的前身更为激进，它反对个体独立的精耕细作和多余的装饰，支持纯粹的功能主义学说。从它的倡导者和关键人物的口号中就可以看出其主旨内涵——阿道夫·路斯（1870—1933）宣称“装饰就是罪恶”，而勒·柯布西耶（1887—1966）认为房屋应该是“居住的机器”。1930年至1933年间担任包豪斯学校校长的路德维希·密斯·凡德罗（1886—1969）给我们留下了简洁的极简主义格言“少即是多”以及“细节就是上帝”。真正符合时代精神的现代建筑在当时成了一种精简而实用的创作，它的审美趣味不是源自附加性装饰，而是其显而易见的适用性和所用材料的真实性。工业化的大规模生产和预加工让出色的设计有了为广大人类提供服务的可能，因此现代主义往往与进步的社会愿景相结合。勒·柯布西耶的职业

生涯颇具启发意义。毫无疑问，他是那个世纪的创造性天才之
一，他的一些小型项目，像萨伏伊别墅（1928—1931）和朗香教 30
堂（1950—1954）等都被公认为20世纪的杰作。然而就像许多自信满满地转行从事城市规划的建筑师一样，他为新建城市形式开出的处方可能会带来灾难性的后果。他建议拆除巴黎市中心的大部分区域，以便实行瓦赞计划（1925），即用一组不美观的网格状摩天大楼来取代五花八门的老街区，在平面图上，每一组建筑都是一模一样的十字形，并被无情地强置于城市表面。万幸的是，这一计划并未付诸实施，但是少数建筑师和规划师采纳了这种大重建的净化理念。现在看来，这一后果十分可怕。1972年，拆除密苏里州圣路易斯市普鲁蒂-艾戈住宅项目的行动通常被认为是一个转折点。该住宅项目不过是16年前才根据理性的现代主义理念建成的，却因为不断增加的社会弊症而臭名昭著。建筑师兼评论家查尔斯·詹克斯认为，这是现代主义梦想的终结。

虽然景观设计师已经在尽力应用功能主义学说，找寻着混凝土、钢铁和玻璃的用途，但依然被现代建筑的迅猛势头所席卷，可以说，这一结果并不令人奇怪。设计师克里斯托弗·唐纳德（1910—1979）是最早热情描绘现代主义的人之一，他出生于加拿大，1928年定居英国。他的著作《现代景观中的花园》是关于现代景观设计学的首部宣言。唐纳德不仅讽刺了维多利亚时代的设计师过于繁杂的装饰和精巧的草本花坛，甚至转头嘲笑了伟大的柯布西耶，因为柯布西耶的许多建筑以田园环境为背景。对唐纳德而言，景观的设计必须与建筑一样，建立在理性和带有目的性的原则上。除了欣赏现代主义建筑，他还热衷于日

本传统建筑和花园设计，他认为这些是通过实用性实现了对美观的追求。他在英格兰设计了两座著名的现代花园，其中一座
31 是给由雷蒙德·麦格拉斯设计、位于萨里郡彻特西圣安山的自家住宅所造的花园；另一座花园是给苏联移民瑟奇·切尔马耶夫的作品、位于萨塞克斯郡哈兰德本特利树林的住宅设计的（均建造于1936年至1937年）。但唐纳德发现英国人很抵触新思维，所以当已经移民美国的包豪斯创始人瓦尔特·格罗皮乌斯（1883—1969）邀请他去哈佛大学设计研究生院教书时，便动身去了美国。最终他留在耶鲁大学任教，在那里，他的兴趣从设计转向城市规划和历史保护。实际上早在1946年，他就开始否定现代主义的教条，并警告说："认为某种建筑或规划在本质上比另一种'更好'是一种危险的谬论。"

哈佛的反叛者

唐纳德最初任教的哈佛大学与景观设计学及其在20世纪向现代主义的转变有着密切的联系。1900年，为了纪念校长的儿子查尔斯·埃利奥特，哈佛大学就开设了这门课程。1893年，埃利奥特成为奥姆斯特德-埃利奥特联合事务所的合伙人，与弗雷德里克·劳·奥姆斯特德及他的侄子、继子约翰·查尔斯·奥姆斯特德（1852—1920）共事，然而此后不久，老奥姆斯特德的健康状况就出现恶化，埃利奥特便成了事务所的领导人，并被正式指定为波士顿公园委员会的景观设计师。在说服委员们为该市的公园系统制定一份全面的规划时，他遇到了困难，这种与日俱增的挫败感可能是导致他在1897年由于脑膜炎英年早逝的原因，那时他年仅37岁。这个由他父亲成立的项目交给

了另一位奥姆斯特德——中央公园设计师的儿子小弗雷德里克·劳·奥姆斯特德。尽管哈佛大学和这门新兴学科的联系已经非常紧密，但到该世纪中叶，景观设计学的教学已经陷入了一种墨守成规的状态。当时，教学重点仍然放在奥姆斯特德对田园景观的看法上，他们普遍认为自然主义设计从一开始就优于 32
所有规则式或明显的人造设计。然而当1937年格罗皮乌斯来到哈佛大学后，他决定将建筑学、景观设计学以及城市和区域规划学三个系合并为设计研究生院，随后一切都发生了改变。景观设计学和建筑学的学生们会在工作室的项目中合作，包豪斯精神就这样开始渗透进景观设计学的课程中。唐纳德就是因为他的前卫思想而被引入哈佛大学，并成为改革的催化剂。三名成人学生加勒特·埃克博（1910—2000）、丹·凯利（1912—2004）和詹姆斯·罗斯（1913—1991）也加入了现代主义事业，他们通常被统称为“哈佛的反叛者”。

罗斯的主职是一名花园设计师，虽然他的风头被同时代的杰出人物盖过去了，但在他位于新泽西州里奇伍德的故居，现在还设立着一座研究中心。他在20世纪30年代为《铅笔制图》所撰写的文章中，抨击了轴对称和如画风格的设计方法，其观点颇具启发性和影响力。丹·凯利对哈佛大学的保守思想非常失望，没有毕业就离开了学校。但在第二次世界大战后，他通过与建筑师埃罗·沙里宁的关系，开始参与纽伦堡正义宫设计。在欧洲期间，他参观了许多史上有名的规则式庭院，在那里他了解了小径、树丛和林荫大道等设计词汇。后来，他与沙里宁合作设计了位于印第安纳州哥伦布市的米勒花园（1957），以现代手法应用其中的某些元素。凯利认识到现代主义花园和古典花园在

本质上没有太大的区别（这同杰里科一样），两者可以成功地融
合在一起。他从现代主义中汲取了精简美学，包括简洁的线条
和清晰的几何图形，但他也很乐意在合适的地方将其与古典对
称结合起来。因此他在设计堪萨斯城纳尔逊·阿特金斯艺术博
物馆的亨利·摩尔雕塑花园（1987—1989）时，以平缓的露台和
33 修剪过的树篱向勒诺特致敬。他还以自己独特的方式设计了贝
聿铭事务所负责的得克萨斯州达拉斯联合银行大厦周围的广场
（1986），在其中设计了263个喷泉和由圆形花岗岩树池组成的
柏树网格。

在三个反叛者中，埃克博对追求现代主义的社会愿景最为积极。1939年，他回到了加利福尼亚州，最初在农场安全管理局工作，帮助移民工人和来自俄克拉何马州以及阿肯色州沙尘暴区的难民设计定居点。他将自己受到的一些包豪斯风格训练带到住宅设计上，不仅满足了基本需求，还增进了社区的愉悦感。在第二次世界大战后，当人们对新住宅的需求不断增加时，这种理念给了他很多帮助。他渴望为工人阶级创造良好的环境，这常常与为富裕客户工作的丘奇形成了对照。埃克博在1945年与罗伯特·罗伊斯顿和爱德华·威廉姆斯建立合作伙伴关系，他们最初与丘奇竞争花园设计的工作，但埃克博对景观设计学的意义和前景有着更深远的理解。他们的事务所承担了更大型的项目，比如校园、林荫道、城市广场以及工业建筑和发电厂的周边环境设计，其中有名的包括弗雷斯诺市中心商业街（1965）、洛杉矶联合银行广场（1964—1968）和新墨西哥大学阿尔伯克基校区开放空间（1962—1978）等。1963年，埃克博成为加州大学伯克利分校景观设计学系主任，1964年成立了知名的EDAW

设计事务所（即埃克博、迪安、奥斯、威廉姆斯四人姓名的首字
母），该事务所后来成为世界上最有影响力的景观和城市设计
公司之一。埃克博的周围都是与他志趣相投的优秀设计师，但
他是其中最擅长宣传他们所肩负的使命的人，在1950年出版的
《为生活的景观》一书中，他成功地做到了这一点，如今，这本书
已经成了经典名作。这是一种将最理论化的学说与最实际的实
践相融合的尝试，表明艺术美可以与社会目标结合起来。他发 34
现了城市环境的碎片化和功能失衡，并为此感到担忧，但他也认
为，规划师和设计师的作用，就是利用好人类最佳的合作本能。
劳伦斯·哈普林（1916—2009）在1940年进入哈佛大学之前，曾
在康奈尔大学农学院学习，他差一点就成了罗斯、凯利和埃克博
的同学，但他也深受格罗皮乌斯、唐纳德和另一位前往美国的包
豪斯前讲师马歇·布劳耶（1902—1981）的影响。在结束为第
二次世界大战服役后，哈普林去了旧金山，在丘奇手下工作。虽
然他和客户们相处融洽，但他发现私人花园的工作存在局限，用
他自己的话来说，就是想要打破“花园的盒子”而投身“更大型
的社区工作”，所以四年之后，他从丘奇的手下离开并建立了自
己的事务所。在他为景观设计学学科发展所做的诸多贡献当
中，最为知名的大概是他设计的那些引人注目的城市公园，特别
是位于俄勒冈州波特兰市的两个项目——爱悦广场（1966）和伊
拉·凯勒水景广场，它们都把哈普林喜欢去散步的那种山间风景
用实物抽象地表现了出来。在华盛顿州的西雅图高速公路公园
（1970）中，他把一条高速公路所造成的城市割裂重新联系起来，
并因此受到了赞誉；在旧金山捷德利广场的城市重建中，他所做
的开创性工作也是如此。20世纪60年代，他受到尊重环境的精神

以及“在土地上轻松生活”的理念驱使，在加利福尼亚州的海岸协助规划了一个名为“海滨农庄”的新社区。哈普林是一位创新家，他寻求着新式的协作，同时开辟了具有创造性的新方法。他与自己的妻子、舞蹈演员兼编导安娜一起开发了一种通过环境来记录动作的“运动乐谱”，并在设计和谱写之间建立了联系。他本着包豪斯精神，力图组建协作团队，把不同学科的见解结合起来。此外，哈普林还是提议让公民参与设计过程的早期倡导者。

在其他地方的现代主义

美国并不是唯一一个产生了现代主义反叛者的国家。在丹
35 麦，G. N.勃兰特（1878—1945）反对学院派美术历史主义，但他受到了英国工艺美术花园的空间清晰度的影响。他在根措夫特设计了有序几何式的玛丽比耶格公墓（1925—1936），并因此被人们铭记。许多景观设计师都曾在勃兰特的工作室当过学生，其中最著名的是C. Th.瑟伦森（1893—1979），他对设计最重要的贡献是将有着丹麦文化的景观元素引入了他的建筑设计中，像是将树篱围成椭圆形的围墙等等，他先后在纳鲁姆私人花园（1948）和海宁安吉尔四世工厂（1956）的雕塑花园中展开实践。丹麦现代主义者的典型态度和美国的一样，那就是将花园和景观视作建筑的附属物，把室外空间作为组成了整体结构的一部分。现代主义在瑞典也风头强劲，它和进步的社会理念结合在了一起。霍尔格·布洛姆（1906—1996）不是景观设计师，而是一名城市规划师，但他在1938年至1971年担任斯德哥尔摩公园的主管期间倡导了一项开明的政策，即公园应该被视作社会的必需品，它对文明生活的意义和配有冷热自来水的房子一样重

要。因此，他认为需要进行规划以便积极地利用它们，从而让公园渗透进整个城市中。景观设计师埃里克·格莱默（1905—1959）帮助布洛姆实现了该目标。在美国产生着重于私人花园的加利福尼亚学派时，社会民主主义的瑞典却产生了以公园设计为主、致力于公共服务的斯德哥尔摩学派。斯德哥尔摩学派避开了规则式和如画风格的教条，转而从地域景观中发现灵感。尽管从视觉上看属于自然主义风格，但它包含了理性的规划、实用的目标和现代的材料。

现代主义建筑是一种对传统的突破，但在试图推翻历史风
格和僵化惯例的同时，它最终也为自己创造了束缚。在20世纪
30年代，国际风格成为当时唯一合适的风格并得到了推广，正
如它的名字那样，国际风格可以在世界任何地方应用，且不受历
史、文化或气候影响。带有玻璃幕墙的钢结构高楼受到国际金 36
融界的青睐，远至曼谷、多伦多、墨尔本以及新加坡等城市的商
务区仿佛都成了曼哈顿的复制品。在这种同质化的命运中，景
观设计学得以幸免，一部分是因为18世纪时那条“向场所中所
蕴含的精神求教”的忠告[1]，但气候、土壤和植被等难以改变的地
区差异的影响也同样造成了这个结果。现代主义从未真正取代
过诸如土壤、水和植物等主要的景观材料，这一时期的很多优秀
理念都被保存了下来，关注材料、强调空间、合理进行场地规划，
以及用优雅的装饰有效地带来美感——这些都是现代主义遗留
下来的积极一面。最重要的是，它保持了景观应该实用的观点，
我们将在下一章中进一步探讨这点。 37

① 1731年，英国诗人亚历山大·蒲柏在写给伯灵顿勋爵的书信中，敦促设计师“向场所中所蕴含的精神求教”，这一忠告对英国如画风格花园产生了重要影响。——编注

第四章

实用与美观

自约翰·狄克逊·亨特提出“三个自然”的概念以来，人们就一直在思考是否需要再增加自然概念的数目。如果一处农业景观依照“野化”政策被故意荒置，那么由此产生的景观，是属于第一自然（荒野）、第二自然（耕地）还是第三自然（带有审美意图而设计的景观）呢？“景观”一词本身就带有自然与文明相混合的含义，有人说我们需要“第四自然”的概念来涵盖拥有复杂概念的区域，比如人工管理的自然保护区、围垦景观、恢复后的栖息地等等。但即使没有第四自然的复杂性，把审美意图当成衡量作为普通景观的第二自然与作为娱乐场所的第三自然的标准的做法仍旧存在很多问题。即便是在日常的场所，美学都常常成为一个争论点。

18世纪在英国地主中流行开来的ferme ornée（字面含义为观赏性农场）很好地说明了这点。这个词语源于英国景观学派的早期代表人物斯蒂芬·斯威策（1682—1745）。观赏性农场是指不仅基于美学原则，而且基于高效农业布置的庄园。其中最著名的例子是诗人威廉·申斯通在什罗普郡利索厄斯的花园，威廉·吉尔平、托马斯·格雷、奥利弗·哥尔德斯密斯、塞缪尔·约翰逊和托马斯·杰斐逊等许多著名人物都曾参观过这座花园。如果可以把一座高产量的农场布置成游乐的场所，那像森林、墓地或水库之类的其他实用场地为什么不行呢？哪怕某些场所在本质上是功利性的，它们为什么不能同时让人心情愉悦，或者至少不丑陋呢？ 38

即便没有这些"改良"，日常的景观也能带来美感。只需回想一下，农田曾多少次作为绘画的主题出现。勃鲁盖尔、霍贝玛、梵高和康斯太勃尔都对田野有着浓厚的兴趣，渴望把它们绘制成画作。只要能产生好的效果，如画派的画家并不反对在画布上挪动景物的位置或夸大垂直距离。如果有必要的话，景观设计师也有办法移动大量的土方。人们雇用景观设计师本来就是为了改善实际的景观外貌，而不仅仅是为了看他们展示方案。不过，向客户、委员会和规划督察展示方案也是一门艺术（也许偶尔并不光彩）。

英国景观设计师协会（现称作景观协会）成立于1929年，当时正值现代主义的鼎盛时期，从业者接受了功能主义的原则，并把它转变成对于实用与美学相结合的关注。景观设计师协会的创立者背景各不相同。杰弗里·杰里科是一名建筑师，在伦敦建筑联盟学院学习期间，他完成了一项对意大利花园的研究。1951年成为协会第一位女性主席的布伦达·科尔文（1897—1981）曾在斯旺利园艺学院学习，她最初打算专攻水果种植。托马斯·夏普（1901—1978）当时是崭露头角的城市规划师，他后来率先提出了都市主义的观点。这个新协会的首任主席是托马斯·莫森（1861—1933），他是著名的花园设计师，后来工作领域延伸到城

39 市规划，还担任过城市规划学会的会长。创立者们对“景观设计师协会”这个名字犹豫了一段时间，最后还是遵循了美国的先例。由于其中很多人从事私家花园的设计，所以他们一度考虑在协会的名字中使用“风景造园师”一词。后来科尔文意识到这险些酿成大错误：“我们可能要花更长的时间才能获得如今这个职业的全部实践机会——如果说我们已经做到了的话。”

这个团体中囊括的许多建筑师和规划师防止了这个新协会成为花园设计师的小圈子，但是“这个职业的全部实践机会”直到第二次世界大战结束后才真正出现，那时候的国民情绪倾向于合作和重建。国家遭到了破坏，回国的军队也希望有更好的生活条件。在这个实行严谨社会主义和凯恩斯主义经济学政策的战后共识时代，景观设计师经常会参与到大型公共项目中。值得注意的是，在20世纪50年代到80年代的英国，公共部门一直是景观设计师的最大雇主。直到玛格丽特·撒切尔新自由主义革命之后，私人单位才开始雇用更多的景观设计师，不过总体上看，基础组织（由众多的当地信托基金构成的慈善机构）至今还是最大的单一雇主。

英国的景观设计师怀念那个社会进步的时代，因为协会的创立者发现了当今已不太常见的目标，他们不仅能够影响大型项目，还可以影响国家的规划政策。而国家面临的问题也很紧迫：新住房的开发、大型基础设施的建设以及农业技术的发展正在迅速改变着景观的面貌。其中的很多问题看上去并不陌生，不仅仅对于那些很容易认为我们如今面临着类似问题的英国读者是这样，对于任何一个生活在正经历着现代化、经济高速发展
40 和景观变化的国家的读者而言都是如此。因此我有必要对这一

时期的英国做更详细的说明。

农　业

农业是一个很好的开头，倘若让人们想象一幅景观画面，许多人都会想到田野、庄园和耕作。布伦达·科尔文在1940年首次出版、1970年再版的《土地与景观》中用一章论述了这一观点。她认为，“人性化且宜居”的景观具有有机的美，但可能因为“政策、用途和习俗”的改变而面临风险。她把当时遇到的问题总结成“郊区扩散”（我们现在称之为“蔓延”，这个词将会在后面几章再次出现）、新建道路和与之相关的“带状发展”，以及农业系统的变化。当然，对于带状发展的恐慌存在阶级维度。建筑师克拉夫·威廉姆斯-埃利斯在其所著的《英国与章鱼》中表达了他的道德愤怒，抨击了参与开发的市场力量；地理学家兼作家约翰·布林克霍夫·杰克逊则在自己的杂志《景观》（发行于1851年至1968年）中歌颂了美国日常的沿路景观，包括商业街、拖车露营地和快餐店。如果我们将两者对比来看会很有意思。

在英国，粮食保障是经历过战时配给制的几代人所关心的问题，但是科尔文也很担心工业化的种植方式，她认为没有必要砍掉树篱来建设高产的农场。她写道：“我们很容易把任何基于景观外在的观点都贬为‘无病呻吟’，并仍然认为实用和美观是对立的。”她主张实用与美观在景观中是“从根本上互补的”。但她并不反对改变，只是这改变需要经过深思熟虑，即她赞成经过精心规划设计之后的改变。因此，科尔文在一个中规中矩的章节中突然提出，就场地的形态而言，“我们可能发现六边形 41
的蜂窝系统可以在边角处为树木和谷仓提供位置，因此更为有用”。虽然我认为这个想法从未流行过，但它表明设计师已经在思考如何将生产效率与景观可能具有的其他优点结合起来。

住　房

如果想要像战后的规划立法一样，控制没有节制的投机性发展，那么必须要全面重建城市现有的劣质住房，并在城郊建设规划良好的新城，从而解决住房短缺的问题。城市规划师埃比尼泽·霍华德（1850—1928）在其著作《明日的田园城市》一书中提出了后面这种发展形式的模型。他主张创造一种新的景观类型，把充分的就业机会和令人愉快的社区等城市生活的最佳方面，与乡村生活中新鲜的空气、明亮的住宅和花园等最好的部分结合起来，这样便避免了两者最糟的特点：城市中污浊的空气与高昂的租金，以及农村中的贫困与失业。这种新的混合体被称作城市-乡村模式。在霍华德看来，可以通过建设自给自足的小城镇来实现这一设想，每座城镇的人口不超过三万五千人，这些城镇里不仅有工作和娱乐机会，还有田野和自然的美景。这些理想之地被称作“田园城市”。最早的两座田园城市是莱奇沃思（建于1903年）和韦林田园城（建于1920年），它们都是伦敦的卫星城。在《新城法》（1946）以及后续相关法案的推动下，这些田园城市的建设反过来激起了新城建设的浪潮。第一波浪潮中建成的11座新城，包括埃塞克斯郡的巴西尔登（1949年命名）、赫特福德郡的赫默尔·亨普斯特德新城（1947）、北安普敦郡的科尔比（1950）以及达勒姆郡的彼得利（1948）。第二波浪潮起于1961年至1964年，也是为了应对住房短缺问题。第三波
42 则发生在1967年至1970年间。人们在苏格兰建设了五座新城，

其中包括了另一名开拓者——建筑师彼得·扬曼（1911—2005）所设计的坎伯诺尔德新城（1956）。

景观设计师从最初就参与其中。杰里科根据田园城市理念的变体为赫默尔·亨普斯特德新城绘制了第一份规划，用他自己的话说，“它不是在田园中的城市，而是在公园中的城市”。他激进的规划因为遭到当地人的抵制而做了修改，但他后来又被邀请设计水景花园（1947），并尝试在其设计中使用了象征主义。他把装饰性的河道与精致的步行桥设计成蛇形，还在眼睛的位置安置了喷泉，嘴巴所在的部位则是小水坝。建筑师、规划师兼景观设计师弗雷德里克·吉伯德（1908—1984）规划了哈罗新城，他对自己的设计充满信心，在那里度过了自己的余生。吉伯德利用现有的地形构建城镇，把新区安排在高地上，并用山谷的空地来分隔。另一位设计先驱希尔维亚·克劳（1901—1997）也参与了赫默尔·亨普斯特德新城和哈罗新城的设计，后来又参与了巴西尔登的景观规划。

随后，新城开发项目公司开始倾向于使用公司内部的景观设计师，而不是聘请外来的顾问。在有远见的领导引领下，景观团队萌生出新颖的理念。特别是在柴郡的沃灵顿，景观设计师对树林和野花草地的内部结构进行了新的开发，将自然栖息地直接引入住宅的花园大门中。设计师更偏爱乡土物种，基本不使用外来的观赏灌木。在20世纪70年代，这一做法被称作“生态学方法”并流行一时。白金汉郡的新城米尔顿·凯恩思新城如同一座城市一般大，这里也采用了宏大的全城景观规划策略，设计师围绕河谷建设了一个带状的公园系统。新城的不同区域 43
依照不同的种植特色来区分，新城中心以欧洲七叶树、红豆杉和月桂为特色，而河谷里的带状公园种满了柳树和山茱萸，被称作斯坦顿贝里的地区则种植酸橙、白桦和山楂。扬曼也参与了这一项目，他建议美式的网格道路应该柔化为随着景观而流动的柔韧网状结构。这并不是说曲线街道一定会让杂乱的郊区变得更能让人接受，但美国的开发商已经扭曲了“城镇景观”的方法，采取了与地形完全无关的曲线型模式。

自20世纪70年代以后，英国就没有再建设过集中规划的新城。设想下在一个践行乐观社会主义、信仰科技、热衷于筑路的时代，私家车仿佛成了拯救者。如今这些开发背后的思想似乎已经过时了，在许多方面也存在着缺陷，但新城理念的拥护者们仍然认为，与放任市场支配相比，这是针对无法满足的新住房需求所能采用的更好的应对方式。现在，新城的说法已经被生态城市所取代，包括中国在内的一些国家已经开始生态城市的建设。即使英国的新城已经建成，它们依然存在争议。当时任城市规划部长刘易斯·西尔金出席在斯蒂夫尼奇举行的公开会议，宣布英国第一座新城的名字时，抗议者们用“独裁者！”的呼声迎接了他，还将火车站的名字改为“西尔金格勒”，以此表示对集中规划制的反感。正如一条新建的高速公路（或者用英国时事——高速铁路为例）一样，一座新城通常也会被当地的居民认为是强加给他们的，就像流星袭来一样受到“欢迎”。景观设计师的任务之一，就是尽量减少这种不请自来的开发所造成的破坏，试着把它们和
44 周围的景观和谐地融合起来。这也是景观设计师首要道德困境的源头所在：一个从业者到底应不应该提供景观方面的建议，让他并不赞成的项目顺利推进？

任何一个正在经历技术现代化的国家都需要新建大量的基

础设施：公路、铁路、机场、水库、水坝、工厂和发电站等等。想把这些都融入景观中，且不破坏土地的历史特征、宜人的景色或古老森林中动植物的多样性等有价值的特点是一项艰巨的工作，但这正是景观设计师自认为有能力承担的任务。我们继续把英国作为研究案例，在许多开发领域，我们都能找到大量例子。

能 源

能源供应在过去是很紧迫的问题，直到现在仍然如此。1963年，首相哈罗德·威尔逊发表了一篇演讲，其中对技术革命的“白热化”拥护使它流传至今。威尔逊所推崇的技术之一就是核能，因为它为人们提供了廉价且易得的电力。但显然，核电站是大型建筑，并且由于需要远离大型的人口中心，同时靠近冷却用的水源，它必须建设在沿海地区，这些地方虽然人口并不密集，却常常具有较高的景观价值。杰里科是第一批为其提供布置方案的景观设计师之一，参与了位于格洛斯特和布里斯托尔中间的塞文河畔奥尔德伯里核电站项目规划。考虑到其规模和所包含的巨大影响，杰里科把该核电站称作“绝对很可怕的外来户”。这个项目选址在乡村地区，由树篱围成了有机图案式的小场地。杰里科承认，在这种建筑上，景观设计师能做的人性化措施并不多，但至少他可以设计出一个结合了四周田野规模与反 45
应堆几何图形的景观，将其与周围环境联系起来。该方案在模型中展示了一系列直线型高地，令人想到了跟杰里科交好的艺术家本·尼克尔森的抽象浮雕画。可惜，由于低估了可用的土壤量，设计并没有依照预期完成。与此同时，扬曼参与了1958年中央电力局在埃塞克斯沿海的赛兹韦尔建设核电站的争议性规划。克劳同样也成为1959年至1965年在雪墩山国家公园建设的特劳斯瓦尼兹核电站的景观顾问。她在整体式建筑四周进行设计，使得建筑与周围景观和谐地融为一体。1958年，她撰写了《能源的景观》，书中不仅涉及发电厂的选址，还探讨了使配电网络对景观视觉影响最小化的途径。她从来没有怀疑过接纳这些基础设施的必要性，还在书的护封上写着："她承认建设大型炼油厂、核反应堆和电网是基本需求。"

水　坝

景观设计师也会被要求减轻新建水坝和水库的影响。克劳参与了英国表面积最大的人工湖——拉特兰湖的设计。拉特兰湖于1976年启用，为人口密集的东米德兰兹地区提供饮用水。她把水库融入起伏平缓的地形中，对附属建筑的选址提出了建议，并针对"水位下降"，即枯水期不自然的驳岸会暴露出来的美学问题制订具体方案。吉伯德则是另一座巨型水库基尔湖的景观顾问。该水库坐落在诺森伯兰，于1982年启用，容量比拉特
46 兰湖要大。建设该水库是因为人们原以为克利夫兰的钢铁和石油化工业将要扩张，虽然预期并没有实现，但它确保了英国北部永远不会出现水资源短缺的问题，如今，它还拥有风景和娱乐价值。当地的很多人反对淹没北泰恩河谷，这可能会让吉伯德在水坝形状、辅助结构的材料，甚至在已完成的土方工程上播种的特定草种等等问题上，对土木工程师施加更多的影响。

森　林

英国最大的人工林位于基尔德湖畔，是20世纪20年代时，

由负责建立木材战略储备的林业委员会种植而成。为了达到这个单一的目标，委员们既没有考虑美观或者生态，也没有时间去考虑开发他们所负责的这块土地的娱乐潜力。他们成排地种植以西特喀云杉为主的外来的针叶树，这些成排的树一直延伸到其管辖范围的边界，常常在地图上呈现出一条条直线。20世纪30年代，当林业委员会想要在英国湖区尝试这种做法时，曾引发过一阵抗议风暴，但直到1963年他们才开始聘请景观设计师来协助规划种植。克劳是他们聘请的首位顾问，她展示了如何通过植物的成块种植与砍伐使其与地形协调，依照自然特征而非所有权来标明边界。1968年出台的《乡村法》要求林业委员会需要“考虑到自然风景保护与乡村舒适性的需求”，从此以后，森林的经营者必须确保其森林不仅仅是成为用材林的储备地，也是能吸引游客的地方，且其物种组成需要更加多样化，以承载更丰富的野生动植物种类。

公 路

任何重大的基础设施提案都可能遭到反对。认为开发是首要需求的中央政府，与寻求捍卫现有景观价值的地方团体和社区之间经常会产生分歧。景观设计师们常常会处在这样的斗争中，他们试着通过提出缓解方案来证明这些项目不一定会损害当地的主要特色。虽然近几十年中，人们越来越重视与现有社区的接触，但在大部分情况下，景观设计师仍然是以技术为主的局外人。在景观设计师参与的所有类别的基础设施中，道路建设大概最能激发他们强烈的情感。科尔文自1955年起任职于主干道景观处理咨询委员会，是英国最早在该领域工作的人之一。

她在《土地与景观》一书中宣传了美国的“公路定制”观念，即修建与现有地形相协调的公路，而不是爆破形成路堑或者将其高架在突兀的路堤上。她还受到了美国风景公园道理念的影响，关注高速公路的“可视域”，即从公路上可以看到的地域范围。在此区域内，景观在美学和环境方面都受到了保护。这个概念非常复杂，因为这是在20世纪早期创造出来的。科尔文还认为，精心设计的道路可以为景观增色。她从司机的体验出发，考虑到他们需要适当的刺激才能保持警觉的诉求。她反对在离高速太近的地方种植树木，以免阳光透过树枝造成令人不快的闪烁效应。目前，负责管理英国战略性路网的高速公路管理局在进行新路线评估和现有道路改善时，依然会听取景观专家的
48 建议。这样的目的仍是使道路与周围的环境相协调，并利用地形和植被来减轻对当地景观特征的不利影响。

美学与道德

当然，如果你是因为美学或者环境的原因反对道路建设，那么景观设计师精心设计的所有方案都不会改变你的想法。同样，在反应堆周围配置任何巧妙的土方工程也都不会让反核抗议者撕掉他们的标语。个别的景观设计师在受邀为军用机场或高速公路的项目提供设计方案时，可能会产生道德危机。一名反对公路建设的学者把景观设计学称作“清粪行业”，因为从业者总是参与收拾别人留下的烂摊子。景观设计师常常会对批评者指出，有争议的方案很可能无论如何都会推进下去，既然如此，有了设计师的帮助会更好一些。但这种论点的缺陷很容易就暴露出来，举一个极端的例子，比如集中营——再多美学和生

态技巧都不可能让这种道德上可憎的东西为人所接受。顺便说一句，这并不是个牵强的案例。正如之前提到过的，德国的许多景观设计师支持第三帝国。其中一名设计师叫作威廉·休伯特，他为海因里希·希姆莱设计了一座日耳曼风格的纪念馆，名为“撒克逊人的树林”，这座纪念馆后来成了党卫军的祭拜场所。战后，他又成功讨取了胜利者的欢心，因此受命在下萨克森州的贝尔根·贝尔森集中营所在地为希姆莱的受害者们设计了一座纪念馆。一些曾参与协调布置希特勒的新高速公路的景观设计师也曾为波兰景观的日耳曼化设计过激进的方案。有时候我们最好记住，景观设计师并不总是站在天使这一边。

另一个可能会被问到的问题是，对于基础设施的隐蔽和掩
饰行为是否存在着欺骗的意味。这个问题是加利福尼亚州的景 49
观设计师罗伯特·塞耶在其著作《灰色世界，绿色心脏：技术、自然和可持续景观》中提出来的。塞耶认为，我们仅仅想要隐藏现有的技术，是因为我们以它为耻。所以我们试着埋藏管道和输电线，遮挡露天矿场，给工厂加上伪装。他说，如果我们拥有引以为豪的环境可持续技术，可能就会想展示它们了。现在的技术恐惧症也会让步于充满喜悦的技术狂热症。这个观点非常有意思，很显然，它涉及文化的转变，比单单通过景观设计学实践所做的任何改变都更为广泛。目前为止，这种转变的迹象仍然寥若晨星。一方面，人们对于堆肥和人工湿地热情高涨，另一方面却在风电厂的选址问题上存在分歧和争议。看来，我们还
是有很长的路要走。 50

第五章

环境学科

浪漫主义者与超验主义者

当代的环保主义可以追根溯源到从前的浪漫主义者身上，他们背弃了工业化的世界，在自然中寻找慰藉和意义。诗人威廉·华兹华斯（1770—1850）的例子可以用来说明这一点，他设计了自己和朋友的花园，其撰写的《湖泊指南》实际上是对文化景观保护的一种变相恳求；他也是一名景观设计师——尽管当时这个职业名称还没被发明出来。华兹华斯是最早发现风景旅游所带来的问题的人之一。他曾经赞美过自己家乡湖区的美丽和幽静，可一旦它变得广为人知，就没有什么能阻止那些有钱人了。这些富人中，有来自曼彻斯特和利物浦等新兴工业城市的实业家和商人，他们在湖区建造大型住宅，破坏了最开始吸引他们的那些美好特色。同样在湖区拥有一套住宅的艺术家、评论家约翰·罗斯金反对了一项穿过该区域的铁路建设方案，他担心如果建设了铁路，这里很快就会出现像“格拉斯米尔周围的小

酒馆和游戏场”。1884年，罗斯金在伦敦学院发表了一场有力
的演讲，宣称他在科尼斯顿的家中观察到了一种新的天气现象，
名叫“瘟疫之风”或“暴风之云”，这种天气源自当时世界上工 51
业化程度最高的城市曼彻斯特。罗斯金后来陷入了疯癫，但我
们很难不将他那混合了气象学与世界末日预言的怪异产物看作
是对当前弊病的预测，即空气污染、全球变暖、气候变化和极端
天气。

浪漫主义影响了包括拉尔夫·沃尔多·爱默生和亨利·戴维·梭罗在内的美国超验主义人士，而他们的作品又孕育了美国早期的环境保护运动。超验主义者追求让日渐科技化和都市化的社会田园化，他们还相信，自然的奇观和壮丽的风景都是神圣的，应该以尊重和敬畏的心来对待。弗雷德里克·劳·奥姆斯特德读了爱默生和梭罗的著作后受到了很大的影响，以至于一位时评家兰斯·纽曼把他称作“超验主义工程师”。奥姆斯特德把超验主义的观念应用到了实践中，除了在城市中建造田园公园外，他还和自然学家约翰·缪尔（1838—1914）一起，为加利福尼亚州约塞米蒂山谷和马里波萨谷巨杉林的红杉树林提供保护。在奥姆斯特德的努力下，环保主义的元素从一开始就被融入景观设计学中。

环保主义

如果只是为了将“环保主义”的标签留给20世纪60年代广泛的哲学、社会和政治运动的话，那么将奥姆斯特德称作原始环保主义者或者自然资源保护主义者也许更恰当一些。另一位原始环保主义的倡导者是出生于丹麦的景观设计师延斯·延森

(1860—1951)，他在芝加哥定居，在成为独立顾问之前供职于城市公园部门。通过观察芝加哥城市扩张，延森发觉美国中西
52 部的原有景观特征面临消失的风险。他对于环境影响设计所做的贡献是自然主义的“草原式花园”，其中采用了本土植物和材料，依靠近距离观察地域性景观形式而建设。他经常将被他称作“草原河流”的湿地特色与聚集人群的“圆形议会空间”融入景观中。1935年，在延森75岁的时候，他在威斯康星州的埃利森贝伊创立了“克利尔因”——一所包含艺术、生态、园艺和哲学等整体课程的学校。

1949年，另一位生活在美国中西部的居民奥尔多·利奥波德出版了《沙乡年鉴》。利奥波德是威斯康星大学麦迪逊分校野生动物管理专业的教授，也是一名林业工作者和野生动植物管理专家。他提出了人类该为土地负责的观点，即“土地伦理学”。这一观点备受推崇，但也时常会引发争论。他是这样表述的：“一个事物，只有在它有助于保护生物共同体的和谐、稳定和美丽的时候，它才是正确的；否则，它就是错误的。”[①]然而，直到生物学家蕾切尔·卡尔森出版《寂静的春天》，公众对环境问题的意识才真正得到了提高。书里指出，用来控制农作物害虫的化学药物会杀死以这些害虫为食的鸟类——春天寂静无声，因为这些鸟儿奄奄一息。

“生态学”，这个曾经从属于生物学类别下一个专业统计分支的词语，很快就开始成为引领整个世界观的旗帜，它让人们意识到自然界的复杂性和依存性，借用利奥波德的一句话，“将人类的角

① 引自《沙乡年鉴》，侯文蕙译，译林出版社2019年版，下同。——译注

色从土地共同体中的征服者转变成其中平等的一员和公民”。

1968年，执行任务的阿波罗8号宇航员所拍摄的照片《地球升起》让人们清楚地感悟到了连通性的意义。在浩瀚的太空中，地球仿佛一颗发光的蓝色弹珠。它看起来很美，但也很脆弱，就像一艘远航中的宇宙飞船，不得不携带上所有的生命维持系统。 53

一年后，任教于宾夕法尼亚大学的苏格兰裔景观设计师伊恩·麦克哈格（1920—2001）出版了该学科有史以来最有影响力的著作——《设计结合自然》，试着把景观设计学放置在科学的基础上。书里指出，把房屋建设在泛洪区或是流沙地的做法很不明智，我们不应违背自然进程，而是要结合自然开展设计。正如我们所见，当代许多关于环境可持续设计的想法都可以追溯到《设计结合自然》。然而，麦克哈格的想法本身则可以溯源到英国景观学派之类强调以移情法对待自然的设计方法，而不是基于试图支配和控制的设计方法。麦克哈格的不同之处在于，他的理论虽然具有宇宙学和形而上学的尺度，却可以提炼成循序渐进的景观规划方法，并从对地质、土壤、气候和水文等方面的详细调查开始着手。

环保主义的诞生，源自对于人类工业及其污染所造成破坏的抗议。早在1991年，环境科学家蒂姆·奥赖尔登就对环保主义者当中的“技术中心论者”和“生态中心论者”做了区分。前一种是乐观主义者，他们认为现有的经济和社会分工能够处理环境问题，然而后一种，包括深层生态学家、盖亚主义者、社群主义者和红绿联盟等等，他们认为有必要重新分配和分散一些权力。这种更为激进的派别在最近的反资本主义示威中又再次出现。无论景观设计师的个人观点如何，很明显的是，作为一个整

体，景观设计学的实践更像是管理类活动。景观设计学的基本信条是人类与自然的关系可以通过规划、设计和管理得到改善，
54 而不是以对世界的变革为先。虽然麦克哈格通常被认为是严谨的生态学思想家，但归根结底，他也主张通过改良而非变革实现人类与自然的最终和解。《设计结合自然》就是这类工作的典范。

生物区域主义与地方特色

许多景观设计师已经接纳了与环保主义类似的生物区域主义概念。这个术语是反文化运动家彼得·伯格在20世纪70年代提出的，20世纪80年代由记者柯克帕特里克·塞尔推广开来，指的是一项和环保主义精神类似，渴望与自然和谐共处的运动，但这项运动却给当地带来了巨大的负担。生物区是通过其物理和环境的特征来定义的，包括土壤、动植物、地形特征和水文等。虽然文化因素也很重要，但生物区并不是以政治或行政边界来作为区分。一些环保主义者似乎将人类视为敌人，但生物区域主义者把人类看作生物区内的居民，并致力于加强人类与地域之间的联系。显然，这种意识形态与发达资本主义试图使各地趋于相似的全球化趋势截然不同。加利福尼亚州的景观设计师罗伯特·塞耶在《生活区域》一书中反思了生物区域主义对日常生活的意义，并探索了在局部范围内重新入住自然界可能带来的社会效益。他的书一半是回忆录，一半是生活方式指南。书中建议我们学着与当地的周围环境产生联结，生活得离土地更近一些，食用当地产的食物并居住在符合本土地域文脉的住宅中。

总而言之，景观设计学一直推崇和赞颂地方特色。我认为

这源于“向场所精神（守护神）求教”的要求。这个要求可以溯源到一个经典的传统，即特定的区域有着属于自己的地方神灵，比如水精和树精，但其真正的含义其实是“重视该场所的现有 55
特质”。在延森看来，使用当地材料和乡土植物就是一条尊重地方特色的和谐设计之路。在荷兰，蒂塞斯强烈批评了当前将人类与自然隔绝的公园设计方法。他强调有必要建设一种新型的公园，使人们意识到当地景观的丰富性和多样性。在他看来，可以通过把乡村的动植物带进城市，让每个人学习和享受来实现这一目标。他在自己位于布卢门达尔的家里推广了“教育花园”的理念，树立了开创性的榜样。后来，阿姆斯特丹一座城郊住宅区阿姆斯特尔芬公园的负责人 J. 兰德韦尔采用了术语“家庭公园”来形容以本土野生植物为主的公园。兰德韦尔创造了一个至今仍旧知名的例子：他建设了一座名为雅各布斯·彼得·蒂塞斯公园的滨水公园，以此来纪念这位开创性的植物学家。家庭公园对世界各地的景观设计师都产生了重要的影响。在曼彻斯特大学任教的艾伦·拉夫将荷兰的本地植物种植理念引入了英国。他的理念在20世纪70到80年代被广泛采纳，尤其是在第四章里介绍的一些英国新城中，人们开始习惯性地提到“生态学方法”。这在很多方面与现代主义一致，人们选用植物不是为了其艳丽的花朵或者闪亮的叶子，而是为了它们在生态系统中所扮演的角色。他们的观点是，本地植物数量丰富、价格低廉，并且容易种植和维护。可以大量种植桤木、柳树、白桦、白蜡树和橡树等本地树种，以营造“结构性林地”。形式特质并不重要，实际上这种方法几乎是反设计的，而且人们相信，这些景观的用户才是最终决定其形式的人。随着这些人造林的生长，它们所

能提供的各种益处也不断增加，其中包括减弱风力、创造休闲机
56 会、维持野生动植物多样性及提供教育材料等，管理林地的成本则逐步降低，修剪整齐的公园里的观赏植物不可能做到这点。许多在这一时期发展起来的技术，包括建设物种丰富的草地和湿地等方式很快成了景观设计中的常见手段，这些丰富的方法使得直白的“生态学方法”理念失去了其大部分的意义。然而，它也预示了当代的许多理念，如绿色基础设施、生态系统服务，甚至是景观都市主义，我们将在后文深入讨论。

景观生态学和生态系统服务

下一个重大发展是20世纪80年代末出现的景观生态学，这是从景观尺度上对于生态学的理解。它强调格局与进程，其中的许多关键概念，如基质、斑块、廊道和镶嵌体等都具有空间性质。比如，一个“斑块”可能指的是一片树林、草地或湿地；一条“廊道”可以是河岸，甚至可以是高速公路的边缘。哈佛大学生态学家理查德·福曼在《土地镶嵌体：景观与区域生态学》一书中猜测：“对于任何景观或其主体，都存在生态系统和土地利用的最佳空间安排，以最大限度提高生态完整性。”简而言之，规划和设计不仅仅是美观或舒适的问题，景观的布局方式可能会对其生态功能产生影响。举个简单的例子，一条贯穿林地的主干道可能会隔绝一片树丛，如果某一特定物种已经出现了数量下降的情况，那么将很难在孤立的斑块上恢复种群。景观生态学科学地解释了为什么我们会觉得人类扩张造成的栖息地碎片化是一件糟糕的事。因此，如今当景观设计师在为一个开发项目
57 提供建议时，他们不会局限于仅仅做一个针对人类美学和便利

性的方案，也会另外考虑到对栖息地和生态系统的影响。他们可以通过自己的设计，力求维持或提升现有的生态系统连接度。幸运的是，许多对物种多样性有利的特征对人类也具有吸引力，像是大型公园、树木繁茂的河岸或是沿废弃铁路线修建的步道等等。除此之外，景观设计师已经精通了转移栖息地的方法。比如可以小心地把长好的树篱挖出来，重新种植到不同的地方，也可以铲起物种丰富的草皮并移植到精心准备的目标场地。这些方法与18世纪景观改善者所采用的技术有着复杂而巧妙的共通点，像他们之中的“万能”布朗，就经常移栽成年树木，为富有的土地所有者创造更宜人的景观。但有些时候他们也会遭到批评，因为移栽到栖息地的树木并不如留在原地的茁壮。

虽然很多环境哲学家坚信自然界拥有内在的权利并试图通过这一点来捍卫它，但看起来，基于人类思想和需求的论点通常更有说服力。这样的论点被称作“人类中心论”，其中包括自然是许多美学和精神满足感的源泉这样的观点。确切地说，更紧迫的想法或许是，如果没有复杂的自然网和各种生物的大量贡献，人类的生存本就无法持续。在生态系统服务的概念中可以找到该论点的一种表述。在许多方面，这一点至少在柏拉图的时代就已经被人类了解。柏拉图在自己的著作《克里底亚篇》中就曾警告过砍伐森林和水土流失的危险，这一警告的现代表述出现在2005年发表的联合国千年生态系统评估中，全球1 300多名科学家参与了这项耗时四年的研究。问题在于，生态系统给予的很多服务似乎是免费的，但赋予它们货币价值就可以使其参与经济计算。尽管一些环保主义者抵制这样的新自由主义 58
思维方式，但这甚至可能是一个市场。比如纽约会替卡茨基尔斯和特拉华流域的供水服务付费，与建设和运转净水厂的成本相比，这是一笔划算的交易。这样的服务非常广泛，从蜂类为作物授粉到食品药品的提供，从固碳、水和空气净化到废弃物的分解，不过也有非物质的利益，比如提供休闲娱乐和振奋精神的场地等等。

生态系统服务的概念对景观设计师和景观规划师而言可能非常重要。在景观设计学的漫长历史中，这一学科一直在努力摆脱这样的观念——毫无疑问，该职业起源于服务上层客户的风景造园——这是个主要与品味和美学相关的学科，因此多余且肤浅。毫无意外的是，景观设计师们从来都没考虑过这些。事实上，对许多人来说，这是一个充满激情的职业，但有关该学科重要性和中心地位的信息有时候却难以传达出去。然而，如果我们可以好好证明生态系统能提供服务，并且这些服务价值惊人；同时，如果我们还能证明它们明确根植于我们所生存的景观中的话，那么景观设计师提供的服务将会受到前所未有的欢迎。

再生设计

加州州立理工大学波莫纳分校的景观设计学系教授约翰·蒂尔曼·莱尔（1934—1998）吸纳了景观可以为人类服务的想法，提出了“再生设计”的理念。莱尔是《环境再生设计：
59 为了可持续发展》一书的作者，也是波莫纳再生研究中心的首席建筑师。在这个研究中心里，一群教师和研究生建立了一个自产食物和能源并自行处理废物的社区，以此证明了人类可以在有限的可用资源范围内生活而不导致环境退化。莱尔对两种生活方式进行了分类，分别称作“退化”和“再生”。退化的生活耗尽了有限的资源，将破坏性的废品回填到大气、湖泊、河流和海洋等自然的“汇”中，这是一个线性过程，一个指向反乌托邦未来的“单向吞吐量系统”。而另一边，再生的生活则通过循环利用的形式提供了持续的能源和材料替代物。莱尔展示了如何改造景观并把它纳入可再生系统：比如可以在含水层上方建造渗滤池以促进水的补给；在入射辐射高的地方放置太阳能集热器；洗衣、洗碗和洗浴等活动产生的灰水可以重复用于作物灌溉等等。莱尔在书中提了许多建议，其中大量“新型技术”（借用他自己的词语）都已经在波莫纳付诸实践。

如今，莱尔搜集的许多技术已经步入了主流的景观设计学实践。可持续排水系统（SuDS）的设计就是一个很好的例子，我们有时也把它称作水敏城市设计。在传统的排水系统中，水通过下水道和排水沟从场地中排走。在许多老旧的城市系统中，污水和雨洪共用同一条管道，一旦排水量超标，就会造成糟糕的结果——排水井盖被冲走，街道上到处都是“令人愉悦”的粪水喷泉。随着全球变暖扰乱了天气模式，暴雨和泛滥的洪水带来的破坏越来越常见。可持续排水系统采用植草沟（较宽的沟渠）和过滤减缓径流，而透水铺装和渗透装置，比如排水坑、碎石排水沟和渗滤池等则有助于水渗入地下，减轻了洪水的风险 60
（图6）。可持续排水系统的原则是尽快在现场将水处理掉，而不是让水沿着管道排到别处。这是许多再生技术的特点，虽然体积不大，却分布广泛。如果20世纪的景观设计师所面临的美学问题是如何协调数量相对较少的巨型水坝和大型发电站，那么，当今的景观设计师所面临的挑战则是如何排布成百上千的风力机和太阳能电池板。

图6　科罗拉多州杜兰戈的这座高尔夫球场在2007年荣获美国景观设计师协会荣誉奖。它包含层级式的人工湿地和洼地，在水汇入现有湿地和溪流之前对其进行收集与净化

有一个概念汇集了本章所探讨的许多思想，尤其是景观生
61 态学、再生设计和生态系统服务，这个概念就是“绿色基础设施
规划”。关于这个概念，我们必须留到后面关于景观规划的章节再广泛讨论，但最基本的概念是，不论是半自然还是人工设计的绿地空间网络，其所带来的效益与路网、下水道系统或电网同等重要。公园绿地、屋顶绿化、村庄广场、沿河堤岸、小区游园和份地（仅仅列举众多类型中的几个）都可以成为绿色基础设施的组成部分。既然这些正是景观设计师经常关注的领域，绿色基
62 础设施能成为该学科目前所热衷的事情，大概也就不足为奇了。

第六章

艺术的空间

是艺术，还是科学？

有几年，我曾经为景观设计学的专业资格课程挑选过硕士生。这段经历让我明白，很难预测谁会成为优秀的景观设计师，也很难根据学生从前获得的成就来推测其潜力。我留意了一些感兴趣的地方和迹象，它们或许能表明学生具有空间思考的能力。当然，如果他们能提供会画画的证据，那就最好不过了。除此之外，攻读本科学位期间学习的科目或课程对于学生的表现也几乎没有帮助。我们招收了很多拥有地理学、建筑学、植物学、生态学、环境科学和园艺学学习经历的人，还有一小部分优秀的艺术生。其中很有趣的一类是具有科学研究背景的学生，他们在上学的时候喜欢艺术，但由于教学大纲的限制而被迫放弃了。这些人常常能成为景观设计学专业的好学生，因为这一专业为他们提供了一条施展才能的完美出路。他们通常在高中的时候都没有听说过这门学科。

英国的学校体系，或许还有世界上多数的教育结构，一般都
63 会强迫学生在艺术和科学领域中做出艰难选择，这个选择常常会限制他们的一生。我们很少看到学生在学习绘画、摄影或平面设计课程的同时还学习生物、物理或地质等“硬科学”。而景观设计学最吸引人的方面就是，它把这种跨学科看作优点。一些景观设计学的从业者是真正的博学人士，至少大部分是通才学者。他们可以读懂生态学家的报告，同样也能看出康斯太勃尔和塞尚画作的意义。社会和政治意识也同样重要，需要在景观设计学课程中讲解。但如果把景观设计学形容成一个兼容并包的大团体的话，这也并不意味景观设计师个体本身没有自己的倾向和偏见。“景观设计学到底是一门艺术，还是一门科学？”这个老生常谈的问题已经在很多研讨会上被提起过了，但人们的意见仍旧存在着分歧。有一些设计师，如麦克哈格，更愿意将景观设计学看作应用生态学；而与此相反，另一部分设计师主要把景观设计学当作一种艺术形式，把设计好的景观作品视为表达含义的载体，将景观视作表达的媒介。

当然，“艺术”一词很难被定义。广义而言，它可以表示“技术”或“工艺”之类。奥姆斯特德常常这样用这个词——参观过伯肯海德公园后，他曾感叹“人们利用艺术从自然中获得了如此多的美”。然而对工程师的科学，奥姆斯特德也提出：“再也没有比用来满足人们对美的需求更有价值的了。到那时，它们不仅被运用到艺术作品上，也被用到美术作品中。”这是个有趣的论述，不只是因为他把工程和科学放在了为艺术服务的地位上，而且奥姆斯特德还清楚地宣告艺术不仅仅是一种技艺，它与美也存在着某些联系。奥姆斯特德深受18世纪英国的美学思想

影响，在当时，风景造园是绘画和诗歌的姊妹艺术。虽然我们经常以美来描述景观，但如果像奥姆斯特德一样把美术与美联系起来，那就不合时宜了。正如艺术评论家阿瑟·丹托所说，现代 64
主义先锋派摒弃了以美感为首要目标的艺术追求，转而以体现意义作为目的——这并不是说艺术不能具有美感，或是美感对于我们的日常生活并不重要（它显然很重要），也不是说景观设计师不该关心这个问题，但是，在考虑景观设计学的艺术可能性时，这或许不是一个最好的出发点。

杰里科的理论

杰弗里·杰里科坚信景观设计学拥有作为一种艺术实践形式的资格和任务，他也注意到了这一点。正如我们在第三章中所看到的，杰里科相信当景观设计学与美术，（对他来说）特别是与绘画艺术相结合的时候是最强大的。对他而言，景观设计学的任务不仅仅是将各种元素排列整齐，也不是清理视觉混乱——这只是对外观体面的追求。景观设计师更高的使命是创造“像绘画一样有意义”的景观。杰里科对景观的意义有着自己的理论，不过现今已经很难找到相信该理论的人了。受分析心理学家荣格的影响，杰里科认为，设计师可以通过在场地中投入时间，激发出一种全人类共有的心理基础——“集体无意识”。随后的设计将会体现普遍原型，并对景观的参观者们产生十分有力，但在很大程度上是无意识的影响。这个理论神秘且无法验证，但它与我们多数人时常体验到的一种感觉相吻合，即某些地方会具有一种强大的存在感或氛围感。

大部分评论家承认景观可以是有意义的，但能在多大程度

上“设计出”意义却常常存在争议。景观设计师劳里·奥林曾
65 于2012年获得过美国总统奥巴马颁发的国家艺术勋章，他的设计代表作包括对纽约布赖恩公园（1992）和哥伦布圆环（2005）的改造。1988年，他撰写了一篇题为《景观设计学中的形式、意义与表现》的文章，因为他意识到自己的学科已经落入了“原教旨主义生态学的重生语言”的控制中。意义又重新变得流行，导师们也会切实要求学生探索隐喻并解释其概念。这转而促使加州大学伯克利分校景观设计学教授马克·特赖贝思考：“景观是否必须有意义？”（这是他在1995年撰写的一篇文章的标题）特赖贝认为，从一开始就试图树立意义常常会适得其反，设计师应该专注于创造给人以愉悦的场地。如果所设计的场地变得知名，那么意义也将随之而来。

野口勇与艺术和设计的界限

虽然艺术和设计常常被放在一起，但它们中间存在区别，大部分从业者也知道自己属于哪一方。一位景观设计师曾告诉我，他并不渴望创造艺术——他的目标是做出“好的设计”。同样，一位在公共场所放置作品的雕塑家也说，他不会按照规定去创作仅仅作为设计而存在的作品。然而，也有一些从业者反对简单的分类，比如日裔美国人野口勇（1904—1988）。野口勇最初跟随康斯坦丁·布兰库希学习雕塑，但到了20世纪30年代，他开始提交关于公共空间和公民纪念碑的方案。由于受到一些知名的花园委托项目的影响，比起被称作艺术家，他更常被人们当作景观设计师，包括在康涅狄格州布卢姆菲尔德的康涅狄格通用人寿保险公司总部的场地设计（1956）、巴黎联合国教科文

组织大楼前的和平花园（1956—1958）以及得克萨斯州休斯顿美术博物馆的卡伦雕塑公园（1984—1986）。野口勇是一名现代主义拥护者，他的作品受到了日本传统的影响，其中最有名的 66
是位于康涅狄格州纽黑文市耶鲁大学贝尼克珍本图书馆内的大理石花园（1960—1964）。花园中有一座低矮的金字塔、一个平衡于一点的立方体和一只竖立的戒指，它们全部由白色大理石制成，园内看不到任何植物，借鉴了以著名的龙安寺花园为代表的禅宗传统枯山水园林而设计。野口还设计了有着柔和轮廓的儿童游乐场，其中在雕塑画廊里使用的器材看上去就和在家里一样。实际上，这些方案里的广场本身就是大型的地面浅浮雕。野口打破了“功能性”设计与自我导向的艺术实践之间的界限，他的作品在这两个领域内都很有影响力，但他还是很难说服官方去建造他设计的任何游乐设施。他在一生中只建起了两处游乐设施，其中一处是1966年与雄谷芳夫在东京附近合作建造的儿童乐园；另一处于1976年开放，2009年修复，坐落在佐治亚州亚特兰大的皮德蒙特公园内（该公园由奥姆斯特德设计）。

有时会存在一种说法，即艺术家以自我提升为追求目标，为自己提出的问题寻求答案，设计师则是在客户需求的背景下用作品来简单回应，并且在整个设计过程中考虑到该设计的最终使用者。这个说法大致正确，但野口的兼收并蓄和综合实践表明，这中间并没有什么严格的区别。艺术与设计之间的界限是可以互相渗透的。

景观与大地艺术

野口对以地球为雕塑媒介充满兴趣，这使他比其所处的时

代领先了一二十年。在20世纪后半叶，对景观设计学影响最大
67 的艺术形式不是杰里科所认为的绘画艺术，而是雕塑，或者说至少是被称作大地艺术、地景艺术或地景作品的特殊潮流。这种潮流出现在20世纪60和70年代，源于概念艺术和极简主义，也是对当时美术馆系统中艺术商品化的一种特殊回应。大地艺术家罗伯特·史密森（1938—1973）、迈克尔·海泽（1944—　）和詹姆斯·蒂雷尔（1943—　）等选择在内华达州、新墨西哥州或亚利桑那州的沙漠等偏远地区创作，以此表达对美术馆的背弃，不过这种拒绝倒并不一定会针对资助他们工作的富有赞助人和基金会。大地艺术和景观设计学有相通之处，两者通常都是“在地性”的，即只能在其所坐落的地点完成创作。和景观设计学的作品一样，大地艺术的作品是对其被创作出来的地点本身的特征和场所精神的反映，通常也用当地的材料来建造。和景观设计学类似的是，大地艺术也可能涉及大规模的土方工程。比如，史密森的《螺旋堤》（1970）就是该类型中最著名的案例之一，它由玄武岩和泥浆建成，位于犹他州大盐湖的岸边，长460米，宽4.6米，如今表面已被盐晶所覆盖。海泽的《双重否定》（1969）是一条宽9米、深达15米的沟渠，跨越了内华达州的一个天然峡谷。一些大地艺术家参与了矿场土地复垦，这是对景观设计学在物理和概念领域的又一入侵。受渥太华硅土公司基金会的委托，海泽在伊利诺伊州的布法罗岩创作了一系列的古冢象征雕塑。这些完成于1985年的作品借鉴了美洲原住民建造土丘的传统，每个作品分别代表该地区的本土生物：鲶鱼、水黾、青蛙、乌龟和蛇。最新的类似作品是建筑师、艺术家、评论家查尔斯·詹克斯为诺森伯兰郡克拉姆灵顿附近一座曾经的露天煤矿

所创作的地形雕塑——诺森伯兰女神（2012）。和露天煤矿相关的艺术家们常常会陷入环境争议，因为有些人会认为他们在 68
协助和教唆破坏性的工业经营，有时从事这方面工作的景观设计师也会受到这样的指责。

大地艺术并没有特别强调自然，但是一些早期的从业者却对生态环境抱有兴趣。艾伦·桑福斯特的作品《时间景观》（1965年至今）由下曼哈顿地区的一块长方形土地组成，艺术家在其中种植了前殖民时代在那里生长的物种。这片区域作为不断成长的林地由城市公园部门管理，被人们当作为了曾经覆盖这座岛屿的森林而建的活纪念碑。牛顿·哈里森和海伦·迈耶·哈里森（通常被称作哈里森夫妇）是生态艺术家的先驱，他们参与了诸如流域恢复、城市更新和气候变化应对等相关工作，这一般被视为规划师或景观设计师等环境专业人士的工作领域。比如，他们最近的装置艺术《温室英国2007—2009》中提到，随着海平面的上升，人类从低洼地区撤离时或许能采用的方法。从业者被人类活动会对地球造成什么后果的伦理问题驱使着，以生态艺术和环境艺术接替了大地艺术。其中一些作品融入了景观设计实践中，在艺术家与景观设计师之间也有过成功的合作。其中最知名的是艺术家乔迪·平托与景观设计师史蒂夫·马蒂诺合作的帕帕戈公园（1992），坐落在亚利桑那州的斯科茨代尔和菲尼克斯的交界处。马蒂诺在这个位于美国西南部的景观作品中率先使用了乡土的耐旱植物。他和平托建造了一个集水结构，从顶部看上去，这个结构就像一棵树的枝丫。通过截留和渗透雨水，该设计有助于场地中特有的树形仙人掌等植
物的再生。 69

景观设计学的先锋派思想？

不管现代主义建筑对哈普林、埃克博和凯利一派的设计师影响如何，哲学家斯特凡妮·罗斯认为，景观设计师和花园设计师已经错过了被其他学科抓住的先锋派潮流。约翰·凯奇的作品《4分33秒》是一首4分33秒的无声乐曲，威廉·伯勒斯的文字剪辑作品可以按照任何顺序重新组合，而像这样的花园作品在哪里呢？这种对作为媒介的材料和过程的内省式关注是先锋派的特征。罗斯试图想象一座先锋派的花园是什么样子——或许它包括了对花园软管的展示？在得出结论之前，花园设计师已经在挑战中退缩了，大地艺术家及其继承者也步入了由此造成的文化真空。当然，对于那些把景观设计学当作设计或规划的人来说，缺乏先锋思想也不是什么问题。

不管怎样，罗斯在其著作《花园有何意义》中的描述过度简化了大地艺术与景观设计学（花园设计）之间的关系。有些花园的案例打破了通常认为的花园模式，其中一些就是景观设计师创造的结果。玛莎·施瓦茨通过在其项目中使用非传统的材料和假植物，震惊了那些持保守观点的人。她以自己的艺术实践为源头，以诙谐的甜甜圈公园（马萨诸塞州波士顿，1979）开启了她的全新职业生涯。公园使用了涂漆的甜甜圈作为家庭花坛（花坛是一种安置在地面上的设计，通常以绿篱围成边界，并用彩色土壤或砾石作为填充）的装饰，以此来取笑法国规则式园林中更宏伟的同类装饰物。当时，这是一项导致该学科分歧的关键作品。施瓦茨为其母亲的复式公寓庭院所设计的斯特拉花园（宾夕法尼亚州巴拉-辛威德，1982）采用了六角细铁丝网、编织网和有机玻璃碎片，这些东西不需要任何园艺技巧就可以维护。她在马萨诸塞州剑桥的怀特黑德研究所屋顶上设计了拼合园（1986），这座花园融合了文艺复兴时期的花园和日本园林的特点，但里面没有一株真实的植物。整齐的篱笆由覆盖着人造草皮的钢铁制成，一块人造的修剪树篱从其中一堵封闭的绿色围墙中水平“生长”出来。一些景观设计师对其嗤之以鼻——这也许是艺术，但它真的属于景观设计学范畴吗？

从这些小项目起步，施瓦茨构建了大型的国际业务机构，赢得了为大城市重要公共空间设计的委托，但她仍然没有失去自己在向行业发起挑衅方面的优势。玛莎·施瓦茨及合伙人事务所对英国曼彻斯特交易广场的重新设计方案（1999年建成）遇到了来自当地政客的麻烦，因为她在其中加入了五株人造棕榈，以此来讽刺该城臭名昭著的灰暗天空和降雨。在定稿方案中，风车取代了这些棕榈树。她为爱尔兰都柏林的大运河广场设计的作品（2008年建成）是斯特拉花园的高档衍生品，上面铺设了由树脂和玻璃制成的红地毯以及一系列倾斜的红色柱子，在夜晚可以发出光亮。施瓦茨的作品通常具有醒目的视觉效果，她似乎征服了设计评审团，但她有时也会因为没有采用更具协商态度的工作方式而遭到批评。由公共空间项目部所维护的耻辱堂网页上展示了一些她备受瞩目的项目，包括英国曼彻斯特交易广场和纽约市中心的雅各布·贾维茨广场（有时也被称作联邦广场或弗利广场）。在后者的设计中，施瓦茨以环绕着绿色半丘的绿色环形长椅为特色，该项目于1997年获得了美国景观设计师协会颁发的荣誉奖。然而，评论家认为对该空间的波普艺术化设计并未给附近写字楼的工作人员提供舒适的场所，在

撰写本文时，迈克尔·范瓦尔肯堡合伙人事务所正在重新改装
71 该广场，说明这一点也许意义重大。但是，单独批评施瓦茨一个人或许太苛刻了。公共空间项目部经常会批评知名的景观设计师，认为他们一些任性的设计只是创造了戏剧性的意象，而不是有生气和活力的城市空间。施瓦茨已经从该学科的“叛逆型天才”变成了老前辈，如今，她与自己的门生克劳德·科米尔共同扮演挑衅者的角色。本书第二章中曾讨论过科米尔的玫红球，他与施瓦茨一样，带有一种玩世不恭的趣味感。他在加拿大安大略四季酒店的项目（2006—2012）中设计了一个高达12米、看起来像一个巨大蛋糕架的铸铁大喷泉，还有一块以花岗岩为原料、铺成阿拉伯式马赛克图案的超大规模“城市地毯”。

共情之法

最强硬的大地艺术可以被视作对景观的一种负担，不过其中也有一个奇特的传统，就是会采取有限且通常脆弱的方式来干预场地。比如英国的理查德·朗（1945—　）和安迪·戈兹沃西（1956—　）以及德国的尼尔斯-乌多（1937—　），他们的很多作品都避免了永久性和纪念性。这样的工作常会受到景观设计师的称赞，因为这似乎与景观设计学实践一样，关注了场所精神或场地特性。在英国利兹都市大学教授景观设计学的特鲁迪·恩特威斯尔也是遵从该模式的在地性艺术家。她走向了与施瓦茨相反的道路——首先接受的是从事景观设计的教育，后来却成为一名艺术家。她称自己的作品“介于大地艺术、雕塑和设计之间，针对特定场地而作，研究了雕塑形式如何与其周围环境相融合，如何与人类活动以及光线、天气、自然生长和衰

退等变化着的元素相互作用”。她的工作并不是为了与景观竞争，而是以某种方式来对其补充。补充的结果不易察觉，属于可能会被偶然发现的附属物（图7）。韩国釜山双年展的《漂流》（2002）和法国吉塞尼的《海浪破碎》（2007）等作品虽然为人们 72 提供了避风之类的临时功能，但功能性并不是其主要目标。

图7　特鲁迪·恩特威斯尔的作品《苹果心》（芬兰图尔库，2008）坐落在名为“树叶上的生命”的院子里。《树叶上的生命》（建于2005年至2009年）是一座奇妙的房屋，由艺术家简–埃里克·安德森创造，其灵感源于自然中的形式。《苹果心》的灵感来源是芬兰的爱情故事《国王与城堡》，这一故事也启发了叶形屋的设计

艺术是可选而非必要

如果我们要问景观设计学能否成为艺术，那很容易就会陷入各种混乱。我们常用赞美的方式来使用“艺术”一词，而不是将其作为一个对人类活动的描述性术语。在褒义语境下，不

可能有不好的或中立的艺术。“建筑”一词也是这样（有时使用首字母“A”），用于标示一种超越了单纯建筑的特殊类别结构。建筑能否被看作艺术，还是说它必须服务的实际目的妨碍了艺术，这仍然是个有待讨论的问题。不管怎样，我们通常都不会用
73 这种方式来评估景观设计学，我认为一部分原因在于“景观设计学”是一个相对较新的术语，还有部分原因是必须归于该学科下的各项活动范围包含了过多的项目，比起创造力，这些项目更需要理性的规划，比如环境影响评估或视觉影响区域评估。我们不会说：“这座公园是景观设计学的杰作，但那座只能算经过了设计的景观。”然而，人们还是有一系列公认的经典杰作，包括龙安寺、兰特庄园、“万能”布朗的公园、中央公园和托马斯·丘奇为唐纳的住宅所设计的花园。这种水平的作品称得上艺术吗？我认为答案是肯定的，但这并不是说景观设计学的动机必须是创造艺术作品。如果特赖贝的观点没错，那么在任何情况下，用寓言和典故来包装设计或许并不是体现其意义的捷径，创造一个激发情感、给予快乐的场所也许才是更为可靠的目标。的确，即便与杰里科相反，仅仅是追求体面，大概也足够让许多从业者
74 和客户满意了。

第七章

服务社会

2008年，玛莎·施瓦茨参与了一个由建筑师凯文·麦克劳德主持的第四频道节目《大城市规划》，这个节目邀请设计师对英格兰北部一座曾经的煤矿城市卡斯尔福德进行空间改造。施瓦茨受邀为市郊的社区新弗莱斯顿设计一片“乡村绿地”。虽然她与当地居民的看法并不一致，但其方案还是付诸实施了。据《园艺周刊》后来发表的一篇文章显示，当地居民给这位美国人在绿地中央布置的雕塑起了个绰号，叫作“玛莎的手指”，以此表达他们对玛莎·施瓦茨的工作方式的感受。参与该项目的另一位景观设计师菲尔·希顿告诉该杂志：“玛莎·施瓦茨是一位相当棒的设计师，但她有些高傲自大……在施加自己的观点之前，设计师需要倾听，这正是玛莎·施瓦茨的错误所在。”另一方面，“委员会式设计”让大多数的景观从业者感到紧张，并会将该词与错误的决策以及令人不快的妥协联系起来。景观**创造者**的远见被那些不愿或不想看到它的人淡化了。这门学科的艺术派人士可能比其他人更敏锐地感受到了这一点，但某个有力的

75 反驳观点也宣称，忽视用户意愿的景观设计就是糟糕的设计。

景观、权力与民主

建筑评论家罗恩·穆尔曾写道："建筑与权力密不可分，它需要权威、金钱和所有权。建造就是对材料、建筑工人、土地、邻居和未来的居民施加权力。"这恐怕是真的，尽管我们可能也会找到关于集体力量的案例，比如美国乡村的谷仓共建活动。在景观设计学领域，哪怕我们仅仅回想下有历史记载的经典设计的话，也能得出类似的结论。这说明建造公园和花园需要宽裕的财富，在大多数情况下，这些都掌握在皇室或特权阶层的手中。这并不一定能让景观的设计者更为轻松，但面对的问题还是存在差异。安德烈·勒诺特在为路易十四布置凡尔赛宫宏大的花园时，必须要面对宫廷的竞争、专制国王的多变想法和王室情妇的干涉，但总体而言，他知道自己的客户到底是谁，也知道如何取悦客户，他不需要太在意其他的什么人。对于凡尔赛宫花园之类经过设计的景观来说，其本质是对于掌握和控制的表达。尽管18世纪的英国风景公园看起来完全不一样，但它们也是财富和权力的展示。权力源于对土地的掌控，财富则用来雇用工人和马匹，以开展必要的河流筑坝和土地重整工作。这些公园大多是给那些经常谈论英国自由的人建造的，但他们所关心的其实是自己在王权专制中所享受的自由，通常不会站在普通人一方。有一件臭名昭著的事情，约瑟夫·戴默，即后来的多尔切斯特伯爵，雇用了"万能"布朗在其米尔顿阿巴斯的庄园工作，他要求布朗把曾经与自己做邻居的村民重新安置到离他的豪宅半英里远的新定居点去。有位顽固的居住者不愿意离开，

戴默竟然命令布朗用洪水将其冲走。在历史上的大部分时间中，景观的设计者唯一要倾听的对象就是他们的出资人。 76

景观设计学的民主化始于19世纪的公园运动。客户变成了公众用户，通常是由民意代表组成的委员会，公园的使用者则是五花八门的市民。提出这一设计概要的部分原因是为了提供一个吸引所有阶层的公园，而在此背后往往寄托着专断的期望，即这种社会的融合可以减轻社会内部的紧张关系。早期现代主义的革命热情推动了事情的进一步发展，将社会任务置于设计事业的核心。德国包豪斯学校（1919—1933）就是围绕着社会主义设计和生产的理想而建立的。许多国家采纳了乌托邦式的观念，即通过预制和大规模生产一种理性、功能性的建筑，可以改善所有人的居住条件。这一观点在英国尤甚。1945年工党政府的首任住房部长奈·贝文宣称，工人配得上最好的待遇。政治家和规划者们寄希望于高层公寓楼，但梦想很快就破灭了，许多英国的塔楼遭到了与第三章中所提到的普鲁蒂-艾戈住宅项目相同的命运。不过，其中也取得了一些重大的成功，比如拉尔夫·厄斯金在纽卡斯尔设计的贝克公寓（1969—1981）就通过对方向和地形的推敲获得了一种场所感。景观设计师参与了被围墙遮挡的低层建筑公共空间的设计和种植，值得注意的是，设计团队咨询了那些即将成为新社会福利房租户的旧贝克排屋居民。

现代主义住宅并不一定得是高层建筑；两名英国建筑师埃里克·莱昂斯和杰弗里·汤森与景观设计师艾弗·坎宁安（1928—2007）合作成立了斯潘建设有限公司，公司在肯特郡、萨里郡和东萨塞克斯郡建造了现代郊区住宅，重新激发了田园城

市运动的理念，并将大型的公共花园融入了房屋前部。斯潘住宅和贝克地产都带有北欧风格，我们很容易发现它们和一些斯
77 堪的纳维亚住宅项目的相似之处，比如丹麦鲍斯韦的南方公园开发项目（1943—1950，由建筑师霍夫和温丁以及景观设计师阿克塞尔·安德森负责）。在该项目中，成排的低层住宅围绕在一片大型社区绿地周围，绿地两侧是高大的杨树。对设计师而言，在所有的这些方案中，他们都有意识地尝试通过开放空间的设计来培养社交能力。

共　情

如果景观设计师想要服务人类——很难想象有在某种程度上不涉及这一点的项目——那就需要共情的能力。这种能力是一种想象力，也是一种无论自己与他人的生活经历和身心状况有多么不同，都能设身处地考虑他人境况的能力。由于通过高等教育获得景观设计师的职业资格需要漫长的周期，且这个职业又属于中产阶级，因此，设计师与他们为之设计的目标人群的生活世界间可能少有重叠。不过，并非大多数景观设计师都是男性。比如据新西兰景观设计师协会的网站显示，虽然在大学中，学习建筑和景观设计学的学生的性别比例起初都是平衡的，但是在注册建筑师中，仅有18%是女性，在景观设计学领域，相应的数字则为42%。近期一本由路易丝·A.莫津戈和琳达·L.朱编写的美国书籍《景观设计学领域的女性：历史与实践论集》中提到，景观设计学为女性提供了一种家庭生活之外的其他选择，而且比起建筑、工程和科学而言，景观设计学对女性从业者的接受程度更高。如果人们对于景观的体验方式有着独特的性别化，那么众多

女性从业者的存在应该能够确保它在设计实践中得到体现，并保证女性的关切能对所创造的场所产生影响。一个正面案例可能 78
是对公园绿地中安全问题的重视和犯罪行为的恐惧。在公园绿地中，一般会避免在人行步道旁种植高大茂密的植株，照明角度也会经过谨慎推敲，公园中还留有替代路线，以便在紧急情况下提供出口。

共情是一件好事，但共情或许也有自己的局限范围。我曾经担任主考的一所景观设计学学校每年都会安排一些时间让健全的学生坐几个小时轮椅，使他们了解如何处理校园里不同表面、水平面的变化和各种坡道。同样，视力正常的学生要戴上眼罩并使用拐杖。最近，据说出于对视障人士的考虑，涌现了一大批感官花园的设计。不过在景观行业中，视觉以及通过绘图交流观点非常重要，以至于我在30年的实践和教学中从未遇到过视障的学生和从业者。对这种形式差异的考量有助于我们理解共情的局限性。我们很容易把自己**认为**别人想要、需要或是应该需要的，而非他们真正想要的施与他们。大部分设计师有时都会犯这类错误。记得我曾经在泰恩-威尔郡盖茨黑德的一处公租房附近，用原木为开放空间设计过一座非常精美的游乐城堡。从图纸上看，这座城堡很棒，建成以后也确实如此，但其幕墙为青少年吸食胶毒提供了理想的隐蔽场所，因此不得不整体拆除。我的失误就在于，我没有询问过任何人他们想要什么。作为一个典型的局外人，我误解了自己所处的环境。如果我住在附近的某条街上，或许早就知道原木城堡是糟糕的想法了。

这座游乐城堡不幸的命运说明了一个最有力的论点，就是设计不仅要考虑到脑海中想象的人，还要考虑真正的人。换句

79 话说，要采取合作或参与的方式。与人交流是获得关于当地知识的方式之一——比如了解叛逆少年的隐藏习惯——这些可能无法通过其他方式获得。如果你能让整个社区，包括被疏远和排斥的那部分人都参与到设计过程中，那就更好了。这样的话，当公园布设好、攀爬架也就位的时候，社区里的人就会把它看作自己的作品，而不再是遥远的权威人士随便强加上去的。这类工作需要包括倾听、建议、解释、谈判、调节、仲裁在内的很多技能，这些技能很少会在设计工作室里传授。此外，耐心与他人合作也会耗费很长时间，因为这是一个涉及大量会议、反馈和图纸修改的迭代过程。对于在设计工作室中接受过系统培训，需要在给定的截止时间前做出最终方案的设计师而言，将其转变成更耐心的渐进过程颇为困难。然而除非付出这样的努力，否则花费在新设施中的时间和金钱就会被浪费了，这种情况在犯罪率高的贫困内城区域尤甚。虽然可以通过强化器械的规格来增强公园长椅和游乐设施的抗破坏能力，但这种防御性的方法也存在局限。鼓励当地团体为自己的周围环境感到自豪是更为有效的办法，因为归属感可以引导社区内部形成各种形式的自我监督。既然如此，那么如何创造这种最佳的归属自豪感呢？

参　与

时间回溯到1969年，一位名叫谢里·阿恩斯坦的美籍规划师发表了一篇题为《市民参与的阶梯》的文章，后来成为经典著作。她指出，参与的程度可以像梯子的梯级一样排列。排在最底层的是“操纵”，这实际上是对参与的一种拙劣模仿。政府官员们通过邀请一些精心挑选过的社区代表来加入委员会，以此

在口头上支持地方民主。这是官员们在“教育”和说服市民，而非市民参与民主。比这略好一点的是“告知”，政府礼貌告知市 80
民那些会对他们产生影响的方案，但信息的传递仍然是单向的。告知当然是走向参与所必需的第一步，但是，除非有回应反馈的机制，否则也不过如此。“咨询”包括通过态度调查、邻里会议和公众听证等方式征求市民的意见。阿恩斯坦提醒，除非与更积极的参与模式相结合，否则这种咨询往往也是虚假的，不过是表面工作而已。在阶梯中处于上层的是“合作关系”，包括政府与社区代表之间通过政策委员会和规划委员会真正分享权力，以及“权力下放”，即通过市民与公职人员的谈判协商使“市民获得对某一特定规划或项目的主导决策权”。阿恩斯坦将“市民控制”排在了最顶层，即把管理和决策的全部权力都让渡给社区的内部人士。当涉及公共资金的时候，执政者通常认为市民全权控制的风险过大，这或许也有道理。询问社区是否具备控制预算和处理场地复杂业务所需的技术与能力也很合理。因此更常见的做法是，以当地政府和社区团体间某些形式的合作关系来寻求参与过程，这种模式往往能成功。

方　法

关于社区参与规划和设计的文章有很多，实际上的确有上百种途径和方法，但很多都是针对问卷调查和公众会议等传统方法的明显缺陷设计的，并没有真正涵盖市民，也没有赋予市民权力。这些方法多种多样，从探索人们想要什么技术，如简报研讨会、未来搜索活动和引导可视化等，到利用当地创造力，如艺术工坊、模型制作和社区地图绘制等方法，再到激励市民参与设 81

计和决策过程的事项及活动本身。与阿恩斯坦阶梯理论中的上层相对应的是地方发展信托基金，这是一系列由社区建立、所有和领导的非营利组织，致力于社会、环境和经济的复兴，通常与其他私人、公共或志愿组织协同工作。或许与景观设计学更相关的一系列方法，是让社区成员（内部人员）与设计者以及生态学家和工程师等其他具有专业知识与技能的人共同联合起来。

专家研讨会与工作小组。“专家研讨会”（charrette）一词源于法语，意为马车或战车。在19世纪，巴黎美术学院建筑系的学生们赶在截止时间前疯狂工作，到了那天，会有一辆推车被推进工作室，学生们必须把自己的设计成果放进去以供老师审阅。这种在截止时间前紧张工作以解决某个问题或完成某个方案的观念同样成了当代的惯例，这也体现了专注的团队合作理念。专家研讨会是短期的团队活动，通常持续数天，参会的专家听取当地利益相关人士（不只是居民，还有其他利益相关的团体，比如政客、赞助者和当地商人等）的意见，尝试为该地构建一个集体愿景，然后在截止时间前疯狂赶工完成绘图。工作小组或设计协助小组与此类似，外部从业者组成小组进驻某地，与利益相关人士合作解决某项问题，或是为了应对某些灾难，如大型工厂关闭、龙卷风袭击或特大洪水。该过程的核心是一个由六到十名专业人员组成的多学科团队，他们将与社区展开四到五天紧张而丰富的合作。一些大学的景观设计学课程会和当地团体举行专家研讨会式的活动，通常与工作室设计项目结合，产出的成
82 果在经过终审后将会提供给社区。

工作坊、设计游戏和情景规划®。这些参与式方法之间有着很强的相似性。这组方法与上一组之间的区别在于，它强调让

人们制订自己的解决方案，而不是只为最终由外部专家所制订的方案做贡献。这些方法通常涉及某种形式的初步定位或可视化阶段。因此在大约30年前，当时的诺丁汉大学社区教育变革小组成员托尼·吉布森博士发明并注册了一个名为“情景规划®”（PFR）的流程。当地人会在该流程中为自己的社区建立一个桌面模型，随后在其社区的不同地点，如图书馆或教堂大厅举办的预告式会议中使用。参与者将意见卡片放置在模型上，表明他们希望看到的变化及增加的项目，比如新的公园或游乐区域、树木的种植、更好的停车场地或地区商店等。随后，卡片可以进行分类和优先性排序，以便为社区制订行动计划，并由工作组后续跟进。设计游戏也与此类似。有时某个用来进行公园规划的版本会包括当地人在公园平面图上放置的缩放纸板图形，这些图形代表了足球场或网球场等各种设施，或是娱乐设备和户外家具模型。这些条目已经提前定好了价格，参与者可以了解在一定预算下哪些是可能实现的，同时加以讨论，并有望就它们的优先级别达成一致。

我将以两个参与式规划和设计应用的案例来结束本章（见方框1和方框2）。

方框1　澳大利亚墨尔本皇家植物园伊恩·波特基金会儿童园，由景观设计师安德鲁·莱德劳设计，2004年建成

儿童园由澳大利亚商人兼慈善家伊恩·波特爵士的基金会资助，一扇以老式园艺工具形状为特色的生锈金属门

83 仿佛在邀请儿童踏入这座神奇的游乐花园。安德鲁·莱德劳是儿童园的首席设计师，他的作品包括植物园的改造项目和学校花园的游乐空间设计。该花园获得了2005年维多利亚旅游奖的最佳新型旅游开发奖，园内包括一条长而曲折的植物隧道、一条蜿蜒的小溪、一堆神秘的废墟和一道被疏花桉（*Eucalyptus pauciflora*）及丛生禾草包围的岩石峡谷（图8）。花园的开发团队包含拥有旅游规划、教育、园艺和艺术等多学科专业知识的成员，但其中也包括了与来自两所不同小学的儿童的合作。这两所小学一所位于市中心，另一所则来自乡村的荒野地区。设计团队参观了学校，收集儿童对于花园的各种想法。设计者会在后续的某个场合向孩子们提供一个概念方案，该方案展示了其中的部分主要元素，如螺旋、隧道和草丘等，随后设计者会邀请他们绘制出其最想要的特色。孩子们也会被带到植物园里，并被鼓励与植物趣味互动。会有一位艺术家与他们一起工作，在自由游戏中完成艺术品创作。通过这些活动，设计师了解到儿童的游戏充满生气和活力，他们喜欢具有空间包围感的场地。回顾其中采用的方法发现，在概念阶段，对设计师而言，儿童园中的交互性活动远比学校中的儿童访谈有帮助。该项目堪称典范，因为它力求与儿童共同建设花园，而不是仅仅为他们开发，同时它采用了积极且有创造性、令人愉快的方法，吸引儿童成为空间的创造者和使用者。虽然花园建设在一座大型植物园内备受瞩目的环境中，但是同样的理念也可以应
84 用到学校操场改造或社区公园与游乐区域的创造中。

图8　澳大利亚墨尔本皇家植物园伊恩·波特基金会儿童园（2004年建成）

85

方框2　英国基础组织

慈善组织“基础”（Groundwork）是英国最大的景观设计师单一雇主。基础组织成立于1982年，是一个以地区环境为行动重点，动员当地人群和资源以改善困难社区前景的组织。这个想法在那些受到煤炭、钢铁和采石业等传统工业衰退影响的地区扎根，那里的社区不仅失去了其主要的就业来源，还要被迫与工业留下的破败景观做斗争。虽然良好的设计始终是一个目标，但该方案也关注解决社会紧张局势，通过培训、教育和工作经验改善人们的生活机会，吸引投资，以及刺激当地经济的发展。该组织还帮助人们思考如何在当地采取行动来对抗全球环境问题。景观设计师同社区开发负责人、青年工作者和项目负责人一道开展工作，与个人和社区团体展开耐心、热情的合作，以此发挥创造力，为社区带来有利改变。基础组织如今已是一个由大约30家非营利信托基金组成的联合体，每年提供数千个项目。选择任何一个项目甚至区域似乎都有失公允，但是为了解释基础组织的活动，我们可以了解一下基础组织利兹分会。在一年内，它帮助年轻人重新设计并翻新了一座滑板公园，与当地居民协会合作管理了一片杂草丛生的林地并对公众开放，将一座荒废的游乐场重新设计成融入自然元素的非规则式游乐场，还在小伦敦小学儿童的参与下，翻新了市中心著名的开放空间维多利亚花园。基础组织的许多项目规模较小，很少能登上光鲜的设计杂志，但这也并不是他们的目标。值得尊重的是

基础组织的许多项目在过去三十多年中对人们的生活产生的累积效应，该方法的大获成功促进了同类组织在美国和日本的成立。 87

第八章

治愈土地

时间回溯到1989年，一位著名的反汽车运动人士给专业杂志写了一封信。当时，英国的许多景观设计师和学生都阅读这一杂志。他的目的是抱怨景观设计师从事道路项目的工作，但他污蔑景观设计学这一学科，把它描述成“清粪行业”，宣称该学科专门清理他人制造的烂摊子，而不是从一开始就在混乱产生前进行阻止。在我的人生中，大部分时间都是在英国东北部生活和工作，从未远离运输煤矿的泰恩河。我目睹了太多由工业造成的混乱，但我认为清理这些烂摊子是一件高尚的事情。环境艺术家米耶尔勒·拉德曼·尤克里斯（1939—　）在开启她在纽约市卫生局的长期进驻艺术家生涯之时，做出的第一件公开举动就是与每一位环卫工人握手，感谢他们为社会所做的重要工作。这类工作通常并不显眼，报酬也很低。尤克里斯建议，应该重新评估这些工作的价值。景观设计师和复垦工程师也值得同样的感谢。当作家J. B. 普里斯特利在1933年到访深陷于经济萧条的英国东北部时，他写道：“我从来没有见过比这里更迫切需要清理的国家。”现在，参观这里的人很难再找到一台卸煤器，而曾经是工业河流的泰恩河的大片河岸如今也草木青葱。88
虽然经常默默无闻，景观设计师在这一转变中却起到了重要的作用，而这不仅仅是一个局部的现象；他们正在世界范围内对后工业景观做着同样的事情。

藏在比喻后面的是什么？

土地复垦是一项高尚的工作，很多我熟悉的景观设计师都说，正是职业中的这部分内容带给了他们最大的满足，但这并不能帮助其避免批评。其中的一种攻击性说法是，这不过是一种表面文章，如同在地毯下清扫灰尘。这种说法也有点道理，因为当景观设计师处理被污染的场地时，其中一个问题就是对有毒物质的处理。如果土壤已经被污染，将其运送到别处也就没有什么意义了，需要就地处理才能解决问题。如果无法减轻毒性，补救的办法是将其堆到场地中的偏远处，用一层防渗的黏土覆盖物包裹起来，然后在顶部铺填土壤并播撒草种。这确实类似于一种糟糕的内务管理实践，通常意味着场地中成为“危险区”的部分永远无法建设，也不能被挖掘出来。尽管如此，考虑到现有的技术，这往往是最有效的解决方案。或许植物修复（利用植物中和毒素的技术）和纳米技术的发展可以为场地污染的处理提供高效而永久的方法。具有讽刺意味的是，杰弗里·杰里科正是在其土地复垦的主体方案进入正轨的时候，决定摒弃将“体面”作为一个充分的目标（收录于他1961年在伦敦当代艺术学院的演讲中，这或许说明了它的主旨）的观点。重工业和制造业在许多西方国家的衰落确保了稳定的佣金流。在经济动荡时

期，废弃场地的产生速度经常比复垦速度要快得多。土地复垦这项工作是必要的，将其贬低成仅仅是清理或追求微不足道的
89 体面似乎很不明智。当代景观设计学在很多方面都反对单纯的布景术。我们也可以采用其他的比喻。如今，讨论“回收利用”废弃土地（棕地），并将该实践与良好的环境管理联系起来已经司空见惯。对于棕地的重新利用，比如用作住宅，可以成为导致城市扩张的农场开发的替代选择。或者我们也可以采用治疗的说法，景观设计师和工程师扮演外科手术团队，来治疗工业对景观造成的创伤。我们也可以援引美术品修复者的形象，他正在修复一件古老杰作多年来所受到的损害。当然，这是看待恢复生态学这种应用科学的一种方式，它常常被应用到复垦项目中，目的是重建在被工业破坏之前就存在的那一类栖息地。

方框3、方框4和方框5提供了这类工作的案例。

方框3　力挽狂澜：英国达勒姆郡海岸（1997—2003）

在达勒姆郡海岸的海滩上，煤渣的倾泻持续了150年。鼎盛时期的煤炭工业每年要倒掉250万吨垃圾，在其运转期间共制造垃圾4 000万吨。该郡臭名昭著的黑色海滩为电影《复仇威龙》和《异形3》提供了荒凉的背景，但是，1993年伊辛顿矿坑的关闭标志着一段肮脏时代的终结，处理采矿造成的严重环境破坏终于有了实现的可能。

1997年，耗资1 000万英镑、由14个组织参与的合作项目“力挽狂澜”开始努力解决这一看起来极为艰巨的任务。项目移走了位于伊辛顿和赫尔登的两座大型废渣堆，避免

其中所含的物质被潮水浸出，成为海滩附近的污染物隐患。机械设备和混凝土塔也被拆除了。同时，扩建沿海步行道， 90
开辟自行车道，并在悬崖顶部和岬角上新建石灰岩草地。这些措施的目的在于重建海岸在工业化之前本有的景观特征。由于在这方面大获成功，该项目于2010年11月被评为英国年度景观，并获得2011年欧洲委员会景观奖第二名。

方框4　德国联邦园艺博览会、园艺节和世博会

有一类活动认为废弃地的复垦是一件值得颂扬的事。德国于1951年恢复了举办国家园艺展的传统，并在这方面走到了领先地位。德国联邦园艺博览会每两年在不同的城市举办一次，成为一种处理战争破坏遗址的机制。在每届德国联邦园艺博览会之后，展出的景观将会被改造成一片永久的公园绿地。首届博览会在汉诺威举行，在本书撰写之时，该展览已经预定于2015年在哈弗尔河流域、2017年在柏林老滕珀尔霍夫机场、2019年在海尔布隆继续举办。

英国政府在1984年至1992年间试行了英国版“联邦园艺博览会”。首届国家园艺节是在利物浦的旧码头上举办的，随后每两年分别在特伦特河畔斯托克、格拉斯哥和盖茨黑德举办，最后一次是于1992年在威尔士埃布韦尔的一座旧钢铁厂里举行。这类展览的目的不仅仅是建设绿地，也是为了吸引外来投资。景观设计常常因政治干预和目标混乱而受到影响，但即便不能每次都在需要振兴的区

域实现经济复兴，这些园艺节也吸引了成千上万的游客，
91 加快了土地复垦的步伐。在世博会和其他大型国际活动中也能发现类似的模式。比如2010年上海世博会的举办地位于黄浦江的两岸，包括了一块面积为14公顷的场地，那里曾是一片钢铁厂和船坞。该设计不仅仅是一项清理工程，中国土人设计的设计团队通过将人工湿地和生态防洪措施结合，实现了利用植物吸收河水中污染物的目的。随后在整个世博会期间，这些水都作为非饮用水得到了利用。

方框5　美国旧金山克里西菲尔德公园（1997—2001）

这座公园与迄今为止提到的其他场地不同，因为它之前不是工业用地，而是军事用地。它曾经是一座在旧金山普雷西迪奥北部滨水区潮汐沼泽上建造的机场。普雷西迪奥是一片占地647公顷的军事建筑群，于1994年被废弃。此处的土壤和地下水受到了航空燃料、杀虫剂和飞机清洁剂的严重污染。

军队完成了初步的清理工作，包括挖掘严重污染的土壤，将其运至场外烧毁，并替代以普雷西迪奥其他区域的原生土壤。污染较轻的土壤则在一个移动的干燥炉中加热，这一过程被称作“低温热脱附”，这种方法可以吸收有机污染物，使得余下的尘土清洁到足以就地掩埋。

哈格里夫斯设计事务所的景观设计师受国家公园管理

局委托，提供了一份符合当地自然和文化历史的规划。在原本平坦的场地上，设计师塑造了模仿和放大风浪作用的雕塑地形。该设计也包括了对潮汐湿地的恢复，随后，观鸟者在湿地中观察到了135种鸟。公园中还有一个备受帆板运动人士欢迎的海滩，在那里可以欣赏到金门大桥的壮观景象。 92

制造场地

哈佛大学设计研究生院教授尼尔·柯克伍德在一本关于土地复垦方法的书中使用了"制造场地"一词作为标题。这个标题一语双关：不仅是由于他描述的这类场地坐落在较老旧的制造业区域，而且这些场地本身也是作为制造业活动偶然的副产品而被造就的。它们如今的特征也要归因于这段历史。更重要的是，要想将这些场地转变成对社会有益的作品，就必须对其重塑。在某些情况下，重塑所需的材料必须要在现场加工，比如清洁的土壤。复垦一般需要两种不同领域专家的合作，其中一方面是场地设计师（景观设计师、规划师和城市设计师），而另一方面是土木和环境工程师。我们也可以将第二个群体扩大，把环境科学家和生态学家包括进来。在将被污染的棕地恢复成有益用地的过程中，没有任何一位专业人员可以独自解决其中的所有问题。作为综合性人才，景观设计师往往能够证明自己是这个合作项目中最有效的协调者。

在这类场地中所遇到的问题可能会令人生畏。有毒物质在其中最难处理，却可能最不明显。诸如重金属、石油和化学残留物质等污染物是肉眼不可见的。在严重的受污染场地中工作的工人必须身着防护服，以免皮肤与有毒物质接触。有些时候，在现场发现的某些原料也有价值。比如一种名为“洗煤”的技术可以用于从矿山废弃物中回收可售煤，这往往会给清理成本带来显著的改变。对受污染土壤进行机械处理的方法正在逐渐被生物技术替代。目前已经发现一些被称为“湿生植物”的植
93 物能够储存大量的水，可以用于治理地下水污染。在犹他州奥格登某座旧输油站开展的一项研究表明，杨树可以阻碍地下水的流动并加快石油的降解，有效防止污染物从场地泄漏。某些超积累植物（例如美国的印度芥菜和向日葵）已被用于在有毒场地吸收重金属污染，有时人们甚至可以收获超积累植物并回收矿物质。植物修复技术似乎是最有前途的土地复垦技术之一，从表面上看，该技术温和且环境友好，但我们还有很多的研究工作要做。比如，如果不收集污染物的话，受污染的植物流入食物链会造成什么后果？这些问题很复杂，需要多学科联合调查才能找到答案。

已封场的垃圾填埋场通常是需要景观设计师关注的一类场地。垃圾填埋场是社会填埋其废弃物的场所，这些废弃物往往被包裹在黏土或塑料制的防渗透外壳中。埋藏了有机物的垃圾填埋场会产生温室气体二氧化碳和甲烷，而后者高度易燃。在会产生气体的垃圾填埋场上一般不允许开发住宅，这不仅是因为甲烷有着火的风险，而且倾倒在填埋场内的材料的沉降可能导致住宅下陷。因此，多数垃圾填埋场都是作为公共开放空间实现再利用的，但它们也带来了其特有的问题。比如过去人们认为在覆盖黏土的填埋场上种植树木会存在问题，因为树根有

可能穿透防水层，但更新的研究显示这并不是一个问题。然而，泄漏的甲烷也会阻碍树木的长势。

尽管存在着障碍，但将垃圾填埋场改造成公园的传统仍然被保留了下来。美国首座此类公园位于弗吉尼亚州的弗吉尼亚海滩（1973年开放），被戏谑地命名为特拉什莫尔山[1]，这是一座68英尺（20.7米）高的人造山丘，由多个清洁的垃圾层在土壤层 94
中夹叠而成，至今仍是深受广大家庭和风筝爱好者喜爱的场所。另一个著名的案例是建于20世纪80年代后期的拜斯比公园，是加利福尼亚州帕洛阿尔托城市垃圾场的一小部分。哈格里夫斯设计事务所与艺术家彼得·奥本海默和彼得·理查德合作创建了一处景观，以微妙的方式感谢了埋在其下60英尺高的废弃物。这座时常有风的海滨公园离机场不远，由于担心植物的根部会造成穿刺，因而没有种植树木。能从顶部燃烧多余甲烷的火炬被纳入了设计方案中，此外公园中还有一片由半埋的电线杆组成的森林。这些火炬和森林是垂直布设的，但随着时间推移，它们会因为下方垃圾的沉降而倾斜。场地内一个“V”字形的大地艺术品算是某种视觉上的幽默。在航空图中，“V”字的意思是“勿在此处着陆”——飞机当然不会，但在冬天，大量的雁类会在飞往温暖南方的途中在此短暂停留。

保留遗迹

有时，土地复垦的评论者会指出，土地复垦可能会破坏当地的传统。生态学家兼牧师约翰·罗德韦尔在写到南约克郡迪恩

① 即“更多垃圾”（trashmore）山。——译注

河谷被抹除的煤矿遗迹时说，对于人类个体，我们会“把记忆丧失看作一种值得关注的病理现象”。人们即使不搬家，也有可能无家可归。如果土地的快速工业化可以被视作一种创伤，维持了几代人生活的工业突然终止也是一种创伤，那么，仓促的复垦计划或许是另一种创伤。然而这并不是必然。20世纪70年代，由美国景观设计师理查德·哈格（1923—　）负责的项目就开辟了一条不同的道路。他说服了正在进行联合湖北岸某个旧煤气厂复垦工作的西雅图市，认为生锈的旧气化装置遗迹没有必要移除。相反，这些令人印象深刻的工业化历史片段被保留下来，成为后来西雅图煤气公园的核心特色。他的案例一直未得
95 到广泛效仿，直到德国拉茨与合伙人事务所在萨尔布吕肯和鲁尔河的后工业景观设计中开启了类似的实践。他们最出名的设计项目是一座曾经的钢铁厂，如今被称作北杜伊斯堡风景公园。公园保留了受到轻微侵蚀的熔炉，花园被布设在旧矿仓里，潜水俱乐部则利用了废弃的储气罐，那里的人工岛礁和沉没的动力游艇如今为水下景观增添了乐趣（图9）。

在德国，这种场地再利用的方式已经几乎成为主流，有很多景观设计师保留而不清除工业遗迹的案例。奥伯豪森规划小组是另一个活跃在鲁尔-埃姆舍尔河区域的景观设计事务所，一直针对关税同盟煤矿和炼焦厂开展工作。这座煤矿的建筑由现代运动建筑师设计，被誉为“世界最美的煤矿”。在1986年关闭之后，煤矿曾面临着被拆除的威胁，但一场运动使其于2001年被联合国教科文组织列入世界遗产名录，并得到了5 000万欧元用于改造的公共资金。荷兰著名的建筑师雷姆·科尔哈斯参与了场地的总体规划，将曾经的洗煤建筑改造成了鲁尔区博物馆。法

图9　20世纪90年代，德国景观设计事务所拉茨与合伙人将鲁尔河畔一座曾经的钢铁厂改造成了著名的北杜伊斯堡风景公园 96

国岱禾景观与城市规划设计事务所也参与了开放空间的设计。
景观设计师同其他设计师和艺术家一道，以参与式方法和当地
学龄儿童合作，从公园着眼，围绕着竖琴形的铁路专线系统等现
有特色设计了公园中的道路和自行车道。虽然巨大的工业综合
体仍然存在，却被占据了建筑间空间的桦树和柳树丛所软化。
设计师围绕着曾经的工业建筑开发了一条漫步道，采用太阳能
照明来活跃夜晚的景色，以巨型花岗岩作品而闻名的艺术家乌
97 尔里希·吕克里姆则在这片后工业森林中建造了一座雕塑公园。

第九章

景观规划

多年以前，在欧洲的某次景观设计学学术会议上，代表们接到了一项任务。他们每位都拿到了一张表格，上面列出了一系列术语——“景观设计学”、“景观设计”、“景观规划”、“园林设计”和“园林艺术”等等——每个术语都用某个任意的几何形状表示。参会代表们被要求各自绘制一幅图表，来表达这些领域之间正确的概念关系。因此，如果用正方形表示园林设计，用圆形表示景观设计学，那么一位认为景观设计学完全包括园林设计的参会代表则会画一个被圆形包围的正方形。如果认为两个领域是重叠的，那就可以用两个交叉的图形来表示。不出所料，绘制出的每张图表都各不相同。

在经过二十多年的国际化讨论和一体化进程后，再重复这个实验也许会很有意思，但我认为，即便是现在也无法做到完全的一致。我希望人们能达成一个共识，即这门学科的总称是“景观设计学”，尽管这个名字并不完美。我还认为大家应该认识到
在这个涵盖性术语下面存在两个互补的活动，即“景观设计”和 98

“景观规划”。两者无疑存在着重叠，但我会尝试对两者做出区分。惯常的做法是绘制一张二元对立的列表，包括如下项目：设计与规划；艺术与科学；小型场地与广阔区域；开创性与问题解决性；综合型与分析型；服务个体与服务社会。这张列表给出了两种活动之间的一些区别。同样毋庸置疑的是，一些景观设计学的学生本能地被该学科偏向艺术的方面所吸引，另一部分则更偏爱分析调查材料、准备方案或撰写报告。然而维也纳工业大学的教授理查德·斯泰尔斯证明了这种二元分割法的不足之处，他注意到景观不能被轻易分割为小型场地和大片土地。这是一个连续体，在连续体的一端是私密的花园式空间，中间是邻里与网络，而另一端是区域景观和整个范围。同样，要说创造性设计不涉及分析思维或解决问题也是不可能的。通常确实是如此。斯泰尔斯指出，规划和设计并不是真正有着不同理论基础的独立活动。设计与规划之间的关系，更像是阴与阳的符号：在设计中总包含一点规划，而在规划中总带有一点设计（这是我的观点，并非斯泰尔斯所说）。当然我可以举一些范例，当景观设计师准备设计花园等某个完全受私人客户掌控的场地时，该活动通常被称为“设计”，虽然它也包含规划的一些方面，比如划定菜圃的最佳位置等。景观设计师在设计公园或公共广场的时候，一般也需要使用类似的技能，即便这类客户已不再是私人个体。代表个人或公司等大型所有者管理大型地产通常是基于理性的活动，但美学可能也是需要考虑的内容——虽然类似活动一般被视作规划。最后，我们举一个经典的景观规划场景，假
99 设某位景观设计师需要代表当地管理部门为其下辖的土地编制一份方案。同样，这在很大程度上似乎也是一个理性决策过程，

但在该过程中经常需要考虑美学和文化，甚至是精神价值。

景观规划的起源

景观规划起源于浪漫主义和超验主义所培育的反城市态度，以及保护自然免受人类侵犯的愿望。正如我们在第五章所看到的，景观设计学之父弗雷德里克·劳·奥姆斯特德与博物学家合作，在美国西部共同保护他们所认定的原始景观。梭罗的名言“世界蕴藏在荒野之中”恰如其分地概括了其背后的哲理。对于梭罗和奥姆斯特德而言，“荒野”与“西部”是同义的，因此保护即意味着对人类活动的排除。正如环境史学家威廉·克罗农所指出的，这种荒野的概念一开始就存在着缺陷，因为它忽视了一个事实，那就是这种所谓的荒野在几个世纪内都曾是美洲原住民的文化景观。不过，这给我们提供了认证景观和受保护景观的概念，特别是国家公园的概念。在美国，这表示某个地点完全无人居住，但英国在第二次世界大战后不久开始首批国家公园认证时，选定的公园都是文化景观，如峰区和湖区，其景观特色和美学特征都是依靠几个世纪的农业活动而形成的。即使“自然美”仍然是一个难以界定的概念，这一理念还是被载入了英国的许多规划立法之中。毕竟，乡村就是自然与人类进程的一个杂交产物。英国有着大量被认证的景观，一些与历史有关，一些与生境匮乏有关，还有一些与文化意义和风景有关，这些都关系到发展规划的起草，为做出在哪里建造什么的决策提供了帮助。 100

国际上也存在着一些认证，比如《拉姆萨尔公约》罗列了全球的重要湿地，认可了其科学、生态和文化价值；不过至少在地位方面位列所有认证之上的，是联合国教科文组织的《世界遗产名录》。被列入这一高端类别的地点必须具有“突出的普遍价值”。该名录的创建理念是保护我们共同的世界遗产，守护雅典卫城等文化瑰宝，或是大峡谷和大堡礁等自然奇观。但后来规则发生了改变，文化景观以及文化与自然的结合体也可以包含在内。一些经过设计的景观，如中国的苏州古典园林被列入了文化遗产；土耳其卡帕多西亚地区的格雷梅峡谷也是如此，那里的民居是在软岩上凿出的，四周环绕着由天然尖岩和塔状岩石构成的壮丽景观。

正如我们在先前的章节中看到的，这种保护自然美景的渴望与改善拥挤城市生活条件的使命相平衡。城市内部开放空间的益处也早已得到认可和推广。费城的建立者威廉·佩恩以其“绿色乡村城镇”的愿景预言了田园城市运动。他躲过了伦敦1665年的黑死病和1666年的大火，所以他希望建筑能够被建设在由开放空间所环绕的大型地块上，这样新城市“再也不会被烧毁，永远保持健康”。需要注意的是，威廉·佩恩的方案提升了安全性，也改善了公共卫生；开放空间可以提供多种不同益处的理念仍是当代思想的核心——用当代的术语而言它是“多功能的”并且“跨越了多重的政策议程”。规划方面的术语可能很艰涩，但它至少是丰富多彩的：公园和花园的总体可以被称为“绿色空间”，永远不会与“棕地”相混淆，但还有一个词叫作“蓝色空间”，是对城市肌理中河流、湖泊、池塘和其他水体的统称。奥姆斯特德的“翡翠项链”是坐落在马萨诸塞州的波士顿与布鲁克莱恩的一系列连
101 环公园，如今被称作“绿色空间网络”，虽然该名称不那么富有诗意，但在本质上是同样的理念。1944年帕特里克·阿伯克龙比的

大伦敦规划是基于对现有景观的系统调查，该规划还建议开辟绿化带以遏制城市发展造成的扩张，并建立以公园、绿地空间以及河流廊道为基础的开放空间系统。

麦克哈格式景观规划——景观适宜度

伊恩·麦克哈格也担心无限制的发展。一些开发项目更适合某些景观而非其他景观，这种想法似乎不过是常识，但当建设在泛洪区的住宅被淹没，或者是构筑在悬崖顶部的酒店坠入大海时，人类的愚蠢程度就变得显而易见了。麦克哈格认为，如果将自然过程和价值纳入思考，我们就可以避免类似的灾害，与自然更和谐地生活下去。他提出了一个考虑到所有因素的方法，被称作“景观适宜度分析”，有时也叫“筛子制图”技术。这项由他开发的技术包含了在醋酸纤维胶片上的信息分层。所以，在考虑某条新建高速的最佳路线时，麦克哈格会将表示基底工程特性的图层与表示高产性土壤、重要野生动植物栖息地和关键文化遗址等的图层叠合起来。当这些图层叠合之后，各种标志最少的干净区域才是建设公路的最佳区域。该方法也适用于区域尺度的发展规划。通常而言，在收集了地理、气候和地质数据后，麦克哈格就可以绘制出适宜度分析图，这类图纸一般划分了农业、林业、休闲和城市开发用地。随着计算机的日益普及，这种依赖于广泛收集和处理数据的方法变得更易施行，“麦克
哈格法”成了GIS（地理信息系统）技术的基础，通过数字化地 102
图层的使用替代了叠加的图纸。景观生态学的出现也丰富了麦克哈格式的景观规划，它提供了一种理论来解释某些生态系统可能出现衰退的原因，并提出了保护和改善这些生态系统的

原则。

通常而言，景观规划不会从一张白纸开始。然而20世纪荷兰圩田景观的开辟是一个常规以外的特例。在这里，新的土地是从海上夺取的，人们可以从头开始，规划农场、堤坝、道路、居住区和林地。这些直线与直线形的平面景观是理性规划的象征，但它们有着自己引人瞩目的美感。然而大部分场所都不是完全依照绘图板上的样式拷贝下来的，多数景观已经发展了几个世纪并产生了分层。“重写本”（罗马石碑或中世纪书卷的名称，其中的部分内容被抹去并重新书写）一词常用来表达一种观点，那就是即便某处景观已经遭到改变，其历史痕迹仍会得到保留。大多数景观规划起步于更复杂的东西，我们甚至不能说这是一个复杂的物体，正如许多理论所指出的，景观是精神的，也是物质的，是主观的，也是客观的。

从特殊景观到整体景观

虽然景观规划起源于对特殊乡村区域的认定和保护，但2000年《欧洲景观公约》（一项欧洲委员会条约，以下简称《公约》）的通过标志着观念的重大转变。《公约》将景观定义为“一个为人所感知的区域，其特征是由自然和（或）人类因素作用与相互作用的结果”，这一定义承认了景观不仅仅是物质的，它还是一种“由人所感知”的东西，换句话说，是被理解和共享的东
103 西。景观被视为“人类周围环境的重要组成部分，是其共有文化和自然遗产多样性的展现，也是其身份认同的基础”。在“适用范围”一章中，《公约》声明它“适用于缔约方的全部领土范围，涵盖自然、乡村、城市和城市边缘地区，包括陆地、内陆水域和海

域；其所涉及的景观可能是突出的，也可能是日常或退化的景观”。虽然这并不代表克罗地亚普利特维采湖群国家公园或法国比利牛斯国家公园等地周围之前的保护认定即将消失，但它的确意味着政治家与规划师必须思考方针，以认定和保存日常景观的品质，让景观更贴近大多数人生存的地域，并提升人们认为在社会、经济、生态和美学方面有缺陷的景观。

《公约》也标志着决策方式从专家决策向普通人决策的重大转变。实际上，《公约》呼吁签字国所做的，是与地方和区域政府在景观政策的确立与施行中共同“建立公众参与程序”。当然这些都存在着不同的解释，毫无疑问，不同国家的实行方式也会不同，但这仍然是一种转变。不管景观规划师认为自己有多么见多识广，他们也不能再依靠自己的单独判断。人们仍然需要专家，但促使公民参与的专业意见将非常宝贵。目前，一场运动正在迫切推动着由联合国支持的《国际景观公约》，这样一来这些思考可能很快就会在全世界应用。

评估任何景观的品质都是一个充满困难的问题。比如试图通过给地图方格中的特征评分以便定量评估的方式最终遭到抛弃，并被一种名为“景观特征评估”（LCA）的方法所取代，至 104
少，在英格兰和苏格兰是如此。景观特征评估是20世纪80年代由土地利用顾问公司开发的，它试图将对景观的描述与可能对其做出的任何评价区分开来。一种被称作“历史景观特征”（HLC）的补充方法增添了“时间深度”这一描述。与《欧洲景观公约》改变特定景观的红线认定一致，历史景观特征特别关注如何保护和管理动态变化的乡村景观。如果我们欣赏景观的原因在于它们是过去变化的重写本，那么从逻辑上而言，我们也必须做好接受进一步变化的准备。问题是，什么样的变化尺度和速度是可以接受的。这里要再次强调，结合公众意见十分重要。

环境影响评估与视觉影响评估

景观设计师的多数规划工作与特定的开发方案相关。他们可以代表开发商申请项目开工前的许可，但也可能代表当地政府，在计划提交后进行评估，或代表反对者，尝试证明某个特定开发项目是有害的。项目的范围变化很大，既有村外田间小屋的开发等小规模项目，也有新机场或高速铁路等大型项目。许多国家采取了一种“环境影响评估”（EIA）程序，强制要求特定类别开发项目的发起人全面审查计划可能造成的任何影响，以及或许能缓解这些影响的所有措施。欧洲环境影响评估相关法案所涵盖的项目种类包括炼油厂、高速公路、化工厂、露天矿场、废弃物处置场和采石场等，不过除此之外，列表中还会有大型集
105 约化家禽养殖场。

环境影响评估包括一项独立但互相有关联的程序，叫作“景观与视觉影响评估”（LVIA），该程序注重评估开发项目对自然景观、视野及视觉舒适度可能造成的影响。景观与视觉影响评估通常由景观设计师完成。当然，对于不需要开展完整环境影响评估的项目，也可以进行景观与视觉影响评估。在开发过程的早期进行这类评估的优点是它们可以充当设计工具，确定避免影响或降低影响程度的途径。之前，景观设计师在做风电厂或新建工厂之类的视觉影响评估时，一般会站在拟建结构场地中，手拿地图和铅笔，尝试绘制出可视区域。增补的部分通过基于地图等高线的绘制工作来完成（图10）。如今，计算机程序可

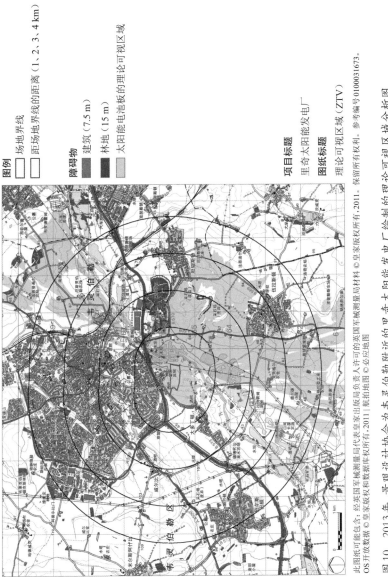

图例

场地界线

距场地界线的距离（1、2、3、4 km）

障碍物

建筑（7.5 m）

林地（15 m）

太阳能电池板的理论可视区域

项目标题

里奇太阳能发电厂

图纸标题

理论可视区域（ZTV）

图10　2013 年，景观设计协会为韦灵伯勒附近的里奇太阳能发电厂绘制的理论可视区域分析图

以更准确地估算出“理论可视区域”（ZTV）。可视化软件也可以提供从特定角度观察拟开发项目外观的可靠图像。有时候问题可以通过缩减开发规模或调整级别来避免。经验表明，在减小影响方面，这些措施往往比种植树木以屏蔽不雅观事物之类的装饰性景观工作更为有效。

绿色基础设施规划

城市绿地的存在很容易被视作理所当然。人们喜爱城市绿地，有时会花费大笔开销以便在其附近居住，但在数次的经济紧缩时期，首先被削减的通常就是对它们的维护，而公共和私人项目也都在蚕食着它们。绿地的案例时不时会被重新阐述，而论述的方式常常反映了该时代的当务之急。在我们所生活的这个时代，经济学家的观点对公共政策拥有巨大的影响力。如果绿
107 地仅仅被视作点缀，那么头脑精明的会计师很可能会得出结论，认为它的维护过于消耗公共资金。因此，我们的观点是需要证明绿地具有实用性和功能性，可以“提供跨越政策议程的交叉利益”，能为我们切实地做些什么。这一想法的最新形式是绿色基础设施规划。

绿色基础设施规划建立在第五章所提到的“环境服务”理念之上。我们提到过的许多公园设施的历史案例很容易就能用这类术语重新分类。奥姆斯特德曾说，中央公园将会成为“城市之肺”，而始建于1878年的波士顿翡翠项链可以被视为一个成功的绿色基础设施项目，通过改良废水处理方式为公共健康带来了益处。当今的观点将生态系统服务划分出一系列方向：有“支持服务”，如土壤的形成，这是所有其他服务的基础；也

有“供给服务”——必需品如食物和燃料的供应；还有“调节服务”，包括大气中碳的捕获等；最后还有包含了自然为人类福祉在各种层面所做贡献的“文化服务”。在文化服务中，景观扮演着重要的角色，提供了美的灵感与享受、历史感与场所感，以及休闲的机会和精神的提升。规划到位的绿色基础设施可以给身处人口稠密城市的居民带来上述益处，还有助于缓解由气候变化引起的一些问题：比如可以通过设计绿地来截留大量洪水并帮助其渗入地下，从而保护建成区。有一个关键理念是“多功能性”，即从水资源管理到栖息地保护，再到促进健康的户外休闲活动，许多不同的功能或活动可以由同一片土地提供。 108

第十章

景观与都市主义

虽然“景观”一词常常被认为是“乡村”或“田园风光”的同义词，但本书中给出的景观设计案例表明，景观设计学既是一种城市实践，也是一种乡村实践。的确，深入而言，景观设计学不只事关农田或田园风光带，它更多的是关于城镇和城市周围所发生的事情。景观的工作一般以某种方式与开发联系在一起，而这大多发生在城市地区。更重要的是，在未来，大部分人会居住在城市中。根据联合国统计，在城市区域居住的世界人口比例已经超过了乡村区域。这一趋势还在持续，因此，到2050年，预计会有70%的人口居住在城镇。越来越多的人将会生活在“特大城市”，即人口超过1 000万的城市群中。东京是目前世界上最大的城市，拥有超过3 400万人口。2000年时，世界上共有16个特大城市，而根据预计，到了2025年将会有27个特大城市，其中21个还会位于欠发达国家。工业革命时期城市快速和大规模的无计划扩张刺激了环境卫生与住房标准的改善，也促进了城市公园的建设。与之相同，当前的城市化浪潮引发了

对于大型居住区中可能提供怎样生活质量的质疑，这些质疑挑 109
战着景观设计学和其他与建成环境有关的学科，促使它们适应特大城市大规模出现的现象。

景观设计学与城市设计

打造宜居城市的任务非常复杂，需要包括景观设计师、建筑师和城市规划师等在内的一系列专业人士的贡献。这些学科中的每一个都具有各自的感受能力，有着各自的训练和教育方法，也有着各自的专业知识。比如，一位景观设计师需要了解哪些树种是最佳的行道树种，并且了解如何在布满下水道、燃气总管和光纤电缆的人行道上种植，而一名城市规划师可能对居住密度和出行方式与建筑形式的关系更有把握。景观设计学给城市问题带来的典型优势之一就是对自然系统和生态的深入理解。在20世纪50年代后期，人们已经越来越确信城市所带来的问题需要专业知识的汇聚来解决，在汽车时代更是如此——因此人们在哈佛大学召开一系列会议，试图为一门名为城市设计的新学科找寻某个共同的根基。这门学科如今已经凭借自身的力量确立了地位，而且拥有大量的大学课程培养该领域的从业者，这些课程通常设立在研究生阶段。不过，与景观设计学和规划不同的是，城市设计并不是一个具有正式认证程序和配套机构设施的行业。虽然这可能是一种解放，但也意味着想要成为一名城市设计师通常需要取得某种相关行业的资格认证，比如景观设计学领域的资格认证。

如果考虑一下城市设计是如何产成的，就会毫不意外地发
现在城市设计和景观设计学之间存在着明显的重叠。我们可以 110

再次回看奥姆斯特德的设计实践，其中很多如今可以被归类为城市设计，正如它们被归类为景观设计学一样。城市社区规划包含了标定和设计公园与开放空间，以及住宅区域、购物中心和交通系统。正如我们所见，奥姆斯特德所设计的公园系统不仅解决了城市环境卫生问题，也为休闲娱乐提供了场所。设计一座公园，必须理解其所处的背景、在城市肌理中的位置、与人们生活和工作地点之间的关系、与街道和其他开放空间的联系以及城市居民和游客可能的使用方式。

景观设计学与城市设计之间的差别在很大程度上是视角问题。在为景观设计学和城市设计专业的学生开办联合工作室的时候，我开始看清这一点。当面对相同的城市场地，通常是已有的开放空间或废弃工厂的复垦土地时，城市设计师的倾向是用建筑来填补，并在建筑之间设计一些零散的小公园及城市广场。而景观设计学专业的学生倾向于使用相反的方式，将少数建筑散布在大片的开放空间中。城市设计师更习惯将绿地视作一种装饰和偶尔的解脱，忽略了大公园及相连绿地系统的功能和生态效益。相反，景观设计师缺乏处理建筑这一形式的信心，在最糟糕的情况下，他们的设计甚至脱离了某座城市。当然这种联合工作室的目的就是为了克服这些狭隘的观念，并且知道其他学科能够提供什么。例如景观设计师需要了解城市发展的经济效益，但是倒不用成为这方面的专家。而城市设计师应该理解绿色基础设
111 施的潜力，不过，他们可以放心将其交予景观设计师来实施。

郊区化、城市蔓延和各种都市主义

交通系统与城市形式之间有着紧密的联系。早在20世纪

30年代的英国，人们就已经对主干道沿线的“带状开发”提出了抗议，这种开发将城镇连接在一起，破坏了乡村的美感。人们注意到，无节制的发展正在吞噬着令人愉悦的风景。在土地资源丰富、汽油廉价的美国，围绕着原本的市中心区域的大型郊区扩张被贴上了“蔓延”的标签。尽管消费者明显偏爱并积极推广郊区的生活方式，这种蔓延还是遭到了大部分建筑师、景观设计师和规划师的谴责，因为它经常与一系列社会和环境弊病相关联。人们认为郊区缺乏传统社区的活力与社交性，同时鼓励人们对汽车的依赖，助长了不健康的生活方式和肥胖的流行。美国也是世界上人均二氧化碳排放量最大的国家，这与低密度的生活和对汽车的热爱有关。托马斯·杰斐逊与威廉·佩恩所倡导的宽敞生活的愿景最终造成了严重的影响。

针对城市蔓延的问题有过很多的对策，其中不少都有着共同的显著特点。最早的对策是新都市主义，是一场通过重新树立有活力的城市领域和强烈的场所意识以对抗社区分散和原子化的城市设计运动。这场运动出现于20世纪80年代，借鉴了建筑师莱昂·克里尔（1946— ）的城市愿景以及理论家克里斯托弗·亚历山大（1936— ）的“模式语言”，两者都提倡回归源于欧洲、历史悠久的城市建设方式。该运动倡导步行化社区，通常以某座公园或城市广场为中心。狭窄的街道会限制交通，其中有些街上还成排种植着树木，而一座宜居城镇的所有组成
部分——学校、托儿所、游乐场和商店——都可以轻易步行到 112
达。在风格上，该运动倾向于保守和回顾性，寻求对传统建筑风格的复制。在新都市主义学派启发下开发的两个最著名的案例是位于英国多切斯特城郊的庞德巴里和美国佛罗里达州的锡赛

德，但评论家们在这些地方发现了某种人造的特征，这或许就是锡赛德被选作电影《楚门的世界》（1998）拍摄地的原因。电影中的主人公毫不知情地过着一种近乎完美的人造生活，而这种生活是电视节目制作团队操控的结果。随后，“理性增长”、“紧凑城市”和“城市集约化”的理念出现了，这些理念都保留了新都市主义关于城市中心步行区的观念，去除了对19世纪欧洲城市的复古式向往。这类设计方法的特征是提供多种住房选择、完善综合的公共交通系统、土地混合利用，以及对农田、城市绿地和具有重要环境意义的栖息地的保护。很明显，景观设计师在这类愿景的实现中起到了重要作用。这种思路在几个欧洲国家，尤其是在英国和荷兰产生了很大的影响。

交通与基础设施

巴塞罗纳、斯特拉斯堡和法兰克福的有轨电车沿着修剪整齐的草坪优雅地从林荫道之间滑过。这些正是高效的公共交通系统与充满魅力的景观设计无缝结合的绝佳案例。景观设计师与工程师合作使交通基础设施充满人性化的例子比比皆是。在瑞典伦德，景观设计师斯文-英瓦尔·安德森（1927—2007）沿着一条铁路线改造了一片线性空间，通过铺设小方石路面和栽植椴树创造出一条步行街。在更大的规模上，荷兰设计事务所West 8已在斯希普霍尔机场的跑道与建筑的四周和中间系统种
113 植了上千株桦树——选择该物种不仅仅是由于其树皮的美丽，也是因为它对雀形目鸟类没有吸引力，因此不会对飞机构成威胁。在许多情况下，交通基础设施也可以成为绿色基础设施的一部分。绿色交通廊道在连接城市基质内部有生态价值的栖息

地斑块方面具有重要意义。有时，道路或铁路会造成分隔，景观设计师可以提供一种补救措施，采用绿色桥梁的形式连接不同栖息地的两端。

在上一段中所举的例子是将交通基础设施中的元素转变为合适场所的精妙介入案例，但交通系统在城市的塑造中有着更广泛、更有战略性的意义。其中的一个知名案例是“伦敦大都会郊区”，它由大都会铁路提供服务和帮助，是20世纪早期在伦敦西北部地区建设的一片狭长郊区。1947年著名的哥本哈根“指状规划”概括出一项战略，依据战略，该城将沿着从密集的城市中心（即“手掌”）所延伸出的五条辐射状通勤列车线（即“手指”）发展，但夹在这五条线中间的是作为农业和休闲用途的楔形绿地。常见的情况是，交通网络的样式、城市的建筑形式和开放空间系统的结构之间具有密切相关性，对这些关联的认识为城市规划提供了工具。城市的发展尽管有设计规范和区划法规的支撑，但仍有很多部分需要由市场决定，这经常让规划师和城市设计师感到懊恼。好在基础设施项目和城市绿地方面的公共资金支出在政治上是可以被接受的，即使对于最奉行资本主义的经济体也不例外，因此其中通常留有为公众利益而塑造城市的余地，或者，至少在基础设施优先于城市发展的社会中就是如此。而对于在特大城市中占很大一部分的棚户区，情况则完全不同，虽然公共资金也可以用于改造这些区域的基础设施、开放空间和服务设施，但它们还是在没有公共资金投入的前提下就出现了。 114

景观都市主义与生态都市主义

正如在哈佛大学的辩论引领了城市设计这一学科的形成一

样，1997年在芝加哥伊利诺伊大学的一场会议宣布了一种名为“景观都市主义”的新**主义**，这是由哈佛大学设计研究生院景观设计学系现任主任查尔斯·瓦尔德海姆所创造的名词。用宾夕法尼亚大学景观设计学教授詹姆斯·科纳的话来说，由于“传统城市设计与规划无法在当代城市有效运转”，这种全新的“思考和行动方式”已然成为必要。这种失败感似乎源于对美国城市持续水平蔓延的思考，源于对发展中国家特大城市发展速度的惊愕，也源于旧工业城市新出现的空心化现象，因为一旦作为其命脉的企业关闭或搬迁，这种现象就可能发生。这一过程以底特律为代表。如今的底特律不再是“汽车城”，而是一片以废弃工厂和破败大酒店的形象为代表的城市景观。景观都市主义者认为，在这些所有的情况下，城市规划师都是无能为力的，而余下唯一能将某座城市联系到一起的，只有它的景观。一种概念化的方式认为，人们的关注点已经从作为城市基本模块的建筑，转移到了作为黏合剂或媒介、将一切联结在一起的景观之中。与城市设计出现的方式类似，景观都市主义学家并未打算设立一种新职业，而是建议整合诸如景观设计学、土木工程、城市规划和建筑等学科的概念领域。景观都市主义方面的硕士课程在北美的数所大学以及伦敦的建筑联盟学院中迅速涌现。

景观都市主义究竟有多大的创新性，以及其对景观设计学传统的成熟理念能实现多大程度的改造一直是讨论的热门话
115 题。景观都市主义的原则之一就是景观如何发挥作用——它为我们做什么——比它的外观更为重要。这与绿色基础设施规划的倡导者所表达的观点非常类似，不过，正如我在书中所论证的，对于功能性的关注是在景观设计学早期出现的一个概念，产

生于奥姆斯特德及其传承者的工作中。景观都市主义学家或许认同这一点，但他们对于奥姆斯特德式传统的争议在于“城市中的乡村”，即将浪漫化的自然融入城市。该观点遭到反对是因为它最多不过是无关紧要的设计，而最严厉的指责认为这是一种伪装或欺骗。他们进一步指出，我们谈论景观与城市两个方面的方式受限于“19世纪的差异与对立镜头”。他们想要主张的是，我们应该消除城市与农村之间的二元区别，并希望人们认识到，城市的足迹远远延伸至我们在传统上所命名的乡村，而后者的组织是为了给城市提供资源，无论是食物，还是饮用水或能源。与此同时，城市内部那些由于与基础设施基本项目相关的工厂或区域的消亡等原因而形成的空间，也向生态演替等自然进程敞开。自从解构主义作为一种文学和哲学运动出现以来，对二元对立的攻击就开始在学术圈流行，但我要指出的是，包括上面这类城乡二元论在内的很多二元论都很有意义，消除乡村与城市之间差异的后果将会是城市蔓延趋势的强化，而且会将城市附近的文化景观也置于危险之中。有时景观都市主义的说辞倾向于“顺其自然”，哪怕这意味着我们的城市将会彻底变得去中心化，呈根茎状网络遍布在景观中。然而正是对“带状开发”的关注推动了英国一系列的规划法的制定和城市绿化带的建设，以此遏制城市的扩张。不加约束的资本主义与放任自流的城市蔓延不一定要占据支配地位。有时，良好的城市规划意 116
味着重新导向、放缓或阻止事情的发生。

从另一方面而言，景观都市主义也有许多意义重大的理念。景观都市主义学家喜欢从长远考虑，他们认识到了场地与城市是随时间不断发展的。科纳在著作中强调了准备“行动的范围”

或“表演的舞台”——这两条短语的含义十分暧昧，既可以指代废弃建筑物的清理等物质性工作，也可以代表更为抽象的活动，如从不同的所有者中收集地块、筹集资金和获得各种许可等等，以便让事情在一定程度上自然发生。景观都市主义推崇灵活的不确定性，以此替代确定的总体规划。景观都市主义学家撰写文章赞扬在底特律闲置土地上涌现的各类城市园艺和农业。那里同时还有些被忽略的场所，如高速公路、管道、污水处理厂、铁路专线和垃圾填埋场之间的边角土地与空隙。巴塞罗那的三一公园（1993）经常被当作参考项目，这座由昂里克·巴特勒和若昂·鲁瓦设计的公园与体育场综合体藏身于一段环形公路的立交桥内。同样著名的是纽约近期建成的高线公园（2005—2010），科纳的菲尔德景观设计事务所同迪勒·斯科菲迪奥与伦弗罗合伙人事务所的建筑师们合作，将曼哈顿一条废弃的货运高架铁路改造成了一座带状公园，其中的植物种植灵感源于在多年的废弃中占据了建筑结构的自播植物（图11）。

我们可以通过记录景观都市主义对于废弃材料的积极利用，为记述其优点的列表再增添一笔。麻省理工学院城市设计和景观设计学副教授艾伦·伯杰在他的著作《废弃地景观》中提出，所有的城市都会产生废弃物，但这些废弃物可以被清除、
117 塑造、平整和重组，实现对社会和环境有益的目的。伯杰写道：“设计师面临的挑战并不是实现没有废弃物的城市化，而是将无法避免的废弃物整合到更灵活的美学和设计策略中。”菲尔德景观设计事务所对弗雷什基尔垃圾填埋场改造（2001—2040）的长期参与被誉为景观都市主义学派实践的楷模，该场所最终将被改造成纽约最大的公园。可以想象，比起土地紧张、由于第二

图11 詹姆斯·科纳的风景园林公司——菲尔德景观设计事务所，与迪勒·斯科菲迪奥与伦弗罗合伙人事务所的建筑师们合作，将曼哈顿一条废弃的货运高架铁路改造成了一座受人喜爱的带状高线公园（2005—2010）

次世界大战后需要处理遭到战争破坏的城市而产生了土地复垦传统的拥挤欧洲，土地供应历来未受到限制的北美洲有必要更努力地推行关于荒地有用性的观点。

景观都市主义是一种蓄意而有益的激励。对景观被描述成“人造场所”的倾听、对消除学科之间界限的讨论、在巨大物理与 118
时间尺度上的思考、对美学重要性的贬低甚至忽略……各种各样的举措已经达到了其预期的效果，激发了实践中的转变、城市问题概念化新方法的提出以及猜测解决方案的新途径。它从未意图取代景观设计学：一个人可以既是景观设计师又是景观都市主义学家，事实上，重要的是那些踏入景观都市主义交会点的人带来了他们各自擅长的知识与技能。但是，作为当下的激进观点，景观都市主义的弧线几近完成。查尔斯·瓦尔德海姆提出，景观都市主义在2010年已经进入了“稳健的中年”，这使美国之外那些刚刚接触到它的人们有些吃惊。2009年，哈佛大学召开了另一场会议，这次的主题是生态都市主义，这是景观都市主义的一个扩展，由设计研究生院的院长穆赫辛·穆斯塔法维提出。在上一个太阳落山之前，世界是否已准备好迎接另一个**主义**，这是一个有待讨论的问题，但新来者保留了其前身所赋予的许多理念，包括人们需要设计类学科来应对所有人都将面临的大范围生态危机。生态都市主义呼吁用新的方法规划未来的城市和改进现有的城市，它似乎已经摒弃了景观都市主义中某些更加尖锐和令人不快的方面，包括其中艰涩的术语。然而很明显的是，景观设计学的价值观念和视角将继续成为这场新运动的中心。在很长一段时间里，景观设计师都会是生态都市主
119 义学家。

索　引

（条目后的数字为原书页码，见本书边码）

E

F

G

L

M

N

O

P

R

S

Ian H. Thompson

LANDSCAPE ARCHITECTURE

A Very Short Introduction

Contents

Preface

Although landscape architecture plays an important role in shaping the everyday places in which many of us live and work, and although it is rooted in practices of manipulating the environment that have a history at least as long as that of architecture or engineering, in many countries it does not enjoy widespread recognition. Why this might be so is one of the questions I will try to answer in this book, but part of the blame must rest with the awkward and misleading disciplinary title, 'landscape architecture'. How we came to be lumbered with this title is disputed. It is often said that Frederick Law Olmsted (1822–1903) and Calvert Vaux (1824–95), the designers of New York's Central Park, were the first to employ the title 'landscape architect', using it on their winning competition entry of 1858, but landscape historian Nina Antonetti recently showed that William Andrews Nesfield, whose elaborately formal designs for the gardens of Buckingham Palace were rejected by Queen Victoria and Prince Albert, described himself as a 'landscape architect' as early as 1849. Other scholars would claim the title of 'first landscape architect' for the designer and horticulturist Andrew Jackson Downing (1815–52), who was one of the first to call for the creation of a large park on Manhattan (Figure 1). There is little doubt, however, that it was Olmsted and Vaux's high profile success that launched the profession. Olmsted is celebrated for his contributions to nature conservation and improved urban

1. Aerial view of Central Park, New York City, originally laid out in accordance with the winning competition entry of 1858 by Frederick Law Olmsted and Calvert Vaux

sanitation, but his greatest legacy consists of the parks he went on to create in numerous American cities, including Boston, MA, Brooklyn, NY, Buffalo, NY, Chicago, IL, Louisville, KY, and Milwaukee, WI. Influenced by English landscape gardening traditions, he believed that he could provide city-dwellers with much needed respite from noise, bustle, and strain by creating pastoral scenery in the midst of the urban environment. Significantly, the winning proposal for the Central Park competition was called the 'Greensward Plan', which included such features as the Ramble, the Sheep Meadow, the Dene, and the Great Lawn.

Despite Olmsted's status as founding father of landscape architecture, he always had misgivings about the name. 'I am all the time bothered with the miserable nomenclature of L.A.', he wrote to his partner Vaux in 1865, 'Landscape is not a good word, Architecture is not; the combination is not—Gardening is worse... The art is not gardening nor is it architecture. What I am doing here in California especially, is neither. It is sylvan art, fine art in distinction from Horticulture, Agriculture, or sylvan useful art... If you are bound to establish this new art, you don't want an old name for it.'

Nevertheless, the name has stuck, despite all the problems it causes. Landscape architects are dogged by persistent misconceptions. The first is that landscape architecture is a sub-discipline of architecture, rather than a separate discipline in its own right; and thus that landscape architects are specialized architects, in the same way that surgeons are specialist doctors. The second is that landscape architects are landscape gardeners (a common slip). Most landscape architects will tell you that at some point they've been invited around by friends to give them 'some advice about the garden'. A former colleague once replied: 'Yes, I'd love to take a look at your garden, but I've got to finish the visual impact assessment for the wind-farm first', and enjoyed the perplexed look that was returned. While landscape architects *do* sometimes design gardens, this amounts to a small fraction of their work. Since landscape architects work, amongst other things, on the layout of business parks, the reclamation of derelict industrial sites, the restoration of historic city parks, and the siting and design of major pieces of infrastructure (such as motorways, dams, power stations and flood defences), the first job of a *Very Short Introduction* to Landscape Architecture must be to answer the question, 'What is it?' Such is the range of work undertaken by contemporary practitioners that Olmsted's idea of a 'sylvan useful art' utterly fails to cover it.

There are various ways this question might be answered, and I will employ a mixture of them all. The first is to take a historical perspective, looking both at the roots of landscape architecture and at the way in which Olmsted and Vaux's sapling discipline grew, developed, and spread. Another angle is to consider the sorts of roles that landscape architects play in contemporary society, the types of commission they undertake, and their relationship to other professionals, such as architects, urban designers, town and country planners, and environmental artists. A third way, and for me the most interesting, is to examine the theoretical bases of the discipline and the various aesthetic, social, and environmental discourses that shape it and distinguish it from cognate fields. The 'What is it?' question then shades into the questions, 'Why do it?' and 'Why is it important?'

As a landscape architect trained in Britain, it is inevitable that I write from the perspective of the Anglophone world (the British and American professions share common roots—and there are strong links to countries in the British Commonwealth, including Australia, Canada, and New Zealand). However, I have tried to temper this with an awareness of the somewhat different origins and perspectives found in other cultures. France, Germany, the Netherlands, and Scandinavia have all played an important role in the development of landscape architecture, but the place where it seems to be growing fastest is currently China. Gardening traditions in Chinese civilization go back at least as far as they do in the West, though the Chinese embrace of landscape architecture is a relatively recent phenomenon, linked to the country's economic take-off since 1979 and the huge physical and social changes implicated in such development. While it is interesting to watch the way that Western ideas of landscape architecture are being grafted onto Chinese culture, we may soon see the influence of Chinese modes of thought and practice having an effect upon the way that landscape architecture is practised in the rest of the world.

There is considerable confusion about terminology, particularly when cross-cultural comparisons are drawn. Uses vary, even between Anglophone countries, and the problem gets thornier when trying to establish corresponding terms in German or French, for example. It would not be difficult to use up the 35,000 words of a VSI on this topic alone, but while I will not be able to avoid some discussion of definitions and nuances of meaning, it will be helpful if I set out what I mean by a few key terms:

Landscape. This is a slippery term, but a useful and widely agreed definition is found in the European Landscape Convention, which states that a landscape is 'an area, as perceived by people, whose character is the result of the action and interaction of natural and/or human factors'. This is helpful because it captures both the idea that a landscape is a tract of land, in other words, something physical, but also that it is something 'perceived by people', which is both of the mind and socially shared.

Landscape architecture. Here is what the International Federation of Landscape Architects says: 'Landscape Architects conduct research and advise on planning, design and stewardship of the outdoor environment and spaces, both within and beyond the built environment, and its conservation and sustainability of development. For the profession of landscape architect, a degree in landscape architecture is required.'

Landscape design. Because the term 'landscape architecture' is flawed, some people prefer the term 'landscape design'. It is a near synonym, but it might be taken to exclude 'landscape planning' (see the next entry). 'Landscape architecture' is the broader term and is also the professional name recognized by the International Labour Organization. In the United States there is a legal distinction between the two terms. Landscape architecture is a registered, state regulated profession, requiring a specific education and successful completion of a registration exam. Landscape design is not state regulated and requires no specific professional academic credentials.

Landscape planning. This helpful definition is offered by the United Nations Education Programme: 'The aspect of the land use planning process that deals with physical, biological, aesthetic, cultural, and historical values and with the relationships and planning between these values, land uses, and the environment.'

To recap, the overarching disciplinary and professional title is 'landscape architecture'. Design and planning can be overlapping activities, and both are aspects of landscape architecture.

List of illustrations

Chapter 1
Origins

The earliest use of the term 'landscape architecture' in print appears to have been in the title of Gilbert Laing Meason's *On The Landscape Architecture of the Great Painters of Italy* in 1828. Meason was a well-connected gentleman-scholar, numbering the best-selling Scottish novelist Sir Walter Scott among his friends, but he had no great following of his own. He used 'landscape architecture' to refer to the setting of buildings in the landscape, rather than to the design of the landscape itself. We might have heard no more about it, had not a fellow Scot, John Claudius Loudon (1783–1843), adopted the expression. Loudon was a prolific designer, writer, and editor, and the founder in 1826 of the influential *Gardener's Magazine*. Many people read Loudon, including his American counterpart Andrew Jackson Downing, whose *A Treatise on the Theory and Practice of Landscape Gardening* went through four editions and sold some 9,000 copies—it included a section headed 'Landscape or rural architecture'. It seems that this was the route by which the term 'landscape architecture' reached the United States and was taken up by Frederick Law Olmsted and Calvert Vaux.

If the expression 'landscape architecture' had not even been coined until 1828, where does that leave my assertion in the Preface that the discipline's origins are as ancient as those of architecture or engineering? Geoffrey and Susan Jellicoe opened their sweeping historical survey of designed landscapes *The Landscape of Man* (first published in 1975) with illustrations showing the alignment of over a thousand menhirs and dolmens at Carnac in Brittany and the arrangement of 50-ton sarsen stones at Stonehenge in Wiltshire, making it clear that mankind has been purposefully reshaping the land since prehistoric times. Books on garden history, similarly, often begin by imagining that the earliest people put up protective barriers around patches of ground, creating the very first yards or gardens. Landscape architecture, as we shall see, is often concerned with the design of functional and productive landscapes, such as farms, forests, and reservoirs, but it shares an interest in aesthetics, pleasure, and amenity with gardening, which links it, not only to the earliest settlements and cultivations, but also to ancient dreams of paradise.

What constitutes paradise has always depended upon the prevailing conditions. For the ancient Persians, enduring harsh conditions on a dusty and riverless plateau, water was manifestly the source of life. They developed underground canals called *qanats* to feed their irrigation ditches, and centred their gardens on intersecting canals, producing the classic quartered design—the *chahar bagh*. The gardens were walled and inward-looking, excluding the desert, and they were filled with all of the things which a desert-living people would enjoy: trees such as date-palms, pomegranates, cherries, and almonds for fruit and shade, cool kiosks, sweet scented shrubs, roses and herbs, pools and bubbling fountains. We derive our word 'paradise' from the Old Iranian (Avestan) word for such exceptional gardens, *pairi-daeza*, which was later shortened to *paridiz*. It is useful to make a distinction between wilderness or 'first nature' and the 'second nature' of human settlement and cultivation. The garden historian John Dixon Hunt has suggested the term 'third nature' for places such as parks and gardens which have been designed with specific aesthetic intent. Landscape architecture, as we shall see, is involved in second nature as well as third nature. Whether there is anything remaining which can be called 'first nature' is a contentious point. Some geologists are already calling our current age the Anthropocene in recognition of the extent of human influence upon the atmosphere and lithosphere. Awareness of the extent of human impact upon the planet should make us all uneasily aware of our collective responsibilities, but it also lends credence to Geoffrey Jellicoe's assertion in *The Landscape of Man* that one day 'landscape design may well be recognised as the most comprehensive of the arts'.

The straight and the curved: formal and informal

Some overview of garden history is needed here because landscape architects are the heirs to centuries of spatial investigation and experimentation conducted by gardeners, and when seeking solutions to new design challenges they often draw upon, or react against, these longstanding traditions. Gardens may be variously classified in terms of their style, but it is useful, from the perspective of design, to locate them on a continuum which has, at one end, formal gardens, which are characterized by geometrical shapes, straight lines, and regularity in plan, and at the other extremity, informal or naturalistic gardens, which are characterized by irregular shapes, curving lines, and much variety. In between these poles are numerous variations and hybrids. The Arts and Crafts style of Edwardian England, for example, was characterized by straight lines, regular geometry, and formality in the plan, but a naturalistic softness in the planting, together with the use of vernacular detailing—employing local materials and traditional construction techniques—in any paving, walls, or other built elements.

History's earliest gardens were mostly formal. Surveying, measuring, and setting-out are, of course, much easier using straight lines and rectilinear shapes. Ancient cities such as Miletus in what is now Turkey or Alexandra in Egypt were built on the same sort of gridiron plan utilized centuries later in a multitude of American cities. Buildings are easier to build from regularly shaped bricks and stones, while the shortest route between two places is in a straight line. It is easier and more efficient to plough a straight furrow than a curving one, and the same can be said for digging canals and drainage ditches. Though human beings tend to describe a slightly curved course when strolling, ceremonial processions are likely to follow a straight line. The landscape historian Norman Newton attributed the origins of the axis, the most potent of spatial ordering devices, to the route of processions through temple grounds. The principal axis of a formal plan is the imagined line which bisects the front elevation of the building at a right angle, be it a temple, a church, or a great house. The axis connects two points and creates the possibility of bilateral symmetry, where one half of the plan mirrors the other. This was characteristic of Renaissance gardens throughout Europe, such as those of the Villa Lante near Bagnaia in Italy or the Luxembourg Gardens in Paris. This way of organizing garden space reached its zenith in the 17th century in the work of André Le Nôtre, master-gardener to Louis XIV of France. At Versailles, 12 miles outside Paris, Le Nôtre created gardens covering an area twice the size of New York City's Central Park. The formality of the plan also extended to the treatment of the plants, which were pruned and clipped until they seemed like green masonry. At gargantuan effort and cost, nature was kept under tight control, though even Louis did not get everything his own way: no matter how many engineers he employed or how many soldiers he ordered to build canals and aqueducts, he never succeeded in getting his fountains to run all day. Versailles became the model for many royal gardens throughout Europe, notably at the Schönbrunn Palace in Vienna, the Peterhof, Peter the Great's Summer Palace outside St. Petersburg, and Hampton Court near London.

In 18th-century England, garden designers and their patrons turned against French formality in favour of designs that became increasingly irregular and naturalistic as the century progressed. There have been various explanations for this change, including the influence of Dutch design on one hand or reports of Chinese traditions on the other. English landowners certainly wished to distance themselves from French formality, which they associated with an abhorrent absolute monarchy. English patrons were often admirers of the landscape paintings of Claude Lorrain and Nicolas Poussin, both of whom had been based in Rome for much of their lives and liked to evoke scenes of Classical Arcadia, taking the landscape of the Roman *Campagna* as their inspiration. More abstractly, the rise of informality in garden design coincides with a growing interest in empiricism. A devotion to rational geometry gave way to careful observation of the apparent irregularities of the natural world. The serpentine 'line of beauty' identified in William Hogarth's *The Analysis of Beauty* much resembles the serpentine curves of a Lancelot 'Capability' Brown lake. In the middle of the century, Brown (1716–83) was ascendant and he remains the best-remembered of his peers, in part because he was so prolific, but also because of his memorable moniker which derives from his habit of telling his patrons, having toured their estates, that he thought he saw 'capabilities' in them, his own word for 'possibilities' or 'potential'. Brown's design formula included the elimination of terraces, balustrades, and all traces of formality; a belt of trees thrown around the park; a river dammed to create a winding lake; and handsome trees dotted through the parkland, either individually or in clumps. Interestingly, Brown did not call himself a landscape gardener. He preferred the terms 'placemaker' and 'improver', which in many ways are conceptually closer to the role of the modern-day landscape architect than 'landscape gardener'. Good examples of Brown's style can be found at Longleat House, Wiltshire, Petworth House in West Sussex, and Temple Newsam park, outside Leeds.

Criticism of Brown began in his own day and intensified after his death. He was criticized in his own time, not for destroying many formal gardens (which he certainly did), but for not going far enough towards nature. Among his detractors were two Hereford squires, Uvedale Price and Richard Payne Knight, both advocates of the new Picturesque style. To count as Picturesque, a view or a design had to be a suitable subject for a painting, but enthusiasts for the new fashion were of the opinion that Brown's landscapes were too boring to qualify. Knight's didactic poem *The Landscape* was directed against Brown, whose interventions, he said, could only create a 'dull, vapid, smooth, and tranquil scene'. What was required was some roughness, shagginess, and variety. This is an argument mirrored in today's opposition between manicured lawns and wildflower meadows. In the United States, where smooth trimmed lawns have been the orthodox treatment for the front yard, often regulated by city ordinances, growing anything other than a well-tended monoculture of grass in front of the house can be controversial.

Humphry Repton (1752–1818), Brown's self-declared successor, argued with the Picturesque enthusiasts, but the public, largely under the influence of the schoolmaster-artist William Gilpin (1724–1804), who published a series of tours to places such as the Wye Valley and the English Lake District, acquired a seemingly unquenchable appetite for Picturesque scenery, and this taste still predominates today. The word 'picturesque', however, has lost much of its original meaning and would seldom be given a capital letter nowadays. For many people it now means little more than 'pretty' or 'attractive'; it has lost its connection with painting.

Repton, however, has a particular place in the genesis of landscape architecture. He was the first practitioner to describe himself as a 'landscape gardener'. He had tried his hand at many occupations—journalist, dramatist, artist, and political agent—before he decided to emulate Brown. He had no deep horticultural knowledge, but he hit upon an ingenious way of presenting his proposals to clients in the form of before-and-after watercolour sketches bound between red covers (Figures 2a and b). By folding out flaps clients could see exactly what changes Repton was suggesting for their estates. These inventive Red Books were the precursors of current methods of visualization—which are more

2a. Panoramic 'before' view, from Humphry Repton's Red Book for Antony House, *c.*1812

2b. Panoramic 'after' view, from Humphry Repton's Red Book for Antony House, *c.*1812

likely to use computer models and fly-through graphics than watercolour drawings. In their own ways, both Brown and the advocates of the Picturesque had imposed their visions upon their clients. Repton was more like a modern-day landscape architect. He understood his clients' needs and listened to what they told him. As a result, he departed from Brown's formula and reintroduced the terrace, close to the house, as a useful garden feature. 'I have discovered that utility must often take the lead of beauty', he wrote, 'and convenience be preferred to picturesque effect, in the neighbourhood of man's habitation.'

'Landscape gardening' becomes 'landscape architecture'

Both Loudon and Downing wrote books with the words 'landscape gardening' in their titles. In the Anglo-American tradition, landscape gardening is regarded as the precursor of landscape architecture. Where the former was a service to a private client, the latter often aspired to be a public service. This change was facilitated by the campaign for the laying out of public parks, particularly in London's East End and the cities of Britain's industrial north. This got going in the 1830s and was part of a movement for social reform which shared the Utilitarian ethos of the philosopher Jeremy Bentham. Loudon was a friend of the philosopher, as were Edwin Chadwick, who campaigned for sanitary reform, and Robert Slaney MP, who argued the case for public parks in Parliament. Utilitarian arguments based on the greatest happiness of the greatest number still underpin many plans and policy decisions in the built environment. 19th-century legislation opened the way in Britain for local authorities to make provision for municipal parks and very soon these became matters of civic pride. The Arboretum in Derby in the English Midlands was one of the first, and Loudon was its designer. It was a gift to the city by a philanthropic textile manufacturer and former mayor, and, as its name suggests, it featured a collection of trees and shrubs, labelled for educational purposes. Parks were supposed to improve people, both physically and morally, while the mixing of different classes in public was thought to promote public order in an age when there was a real fear of imminent revolution. Loudon abandoned his Picturesque enthusiasm in favour of a deliberately artificial approach to layout and planting which he named the 'Gardenesque School of

Landscape'. It featured geometric planting beds and exotic plants raised in glasshouses and then planted out. The Gardenesque would soon become the approved style for Victorian parks, offering copious opportunities for the display of horticultural excellence. Loudon's great contemporary was Joseph Paxton (1803–65), a polymath who rose from being a lowly gardener at the Royal Horticultural Society's gardens at Chiswick House, London, to becoming the celebrated designer of the Crystal Palace at the Great Exhibition of 1851. He undertook the design of many public parks, but one in particular, Birkenhead Park on Merseyside, was pivotal because Olmsted saw it on his visit to Britain in 1850 and it inspired his design for Central Park. If it were not anachronistic, there would be no great difficulty in calling designers such as Loudon and Paxton 'landscape architects', but the term had not been coined in their time.

Meanwhile the job title 'landscape gardener' has not faded away, though it is probably more fashionable and lucrative nowadays to describe oneself as a 'garden designer'. Landscape architecture, as I will show, is by far the broader field, though garden designers, like celebrity chefs, may be better known to the public. The garden now stands in relation to the landscape architect as the private house does to the architect. Just as architects sometimes design private houses, so do landscape architects at times design private gardens—and exhibits at the Chelsea Garden Show—but their bread-and-butter work, particularly in larger offices, is in bigger projects and, as we shall see, many of these are tied, in one way or another, to development. As a profession, landscape architecture only became officially constituted in 1899 when the American Society of Landscape Architects was formed at a meeting in New York. Interestingly they excluded contractors, builders, and nursery-men from their ranks; they did allow in Beatrix Farrand, who designed the celebrated gardens at Dumbarton Oaks in Washington, DC, but she persevered in calling herself a 'landscape gardener' to the end of her distinguished career. In Britain, the Institute of Landscape Architects (now the Landscape Institute) was not formed until 1929, 71 years after Olmsted and Vaux's competition entry had introduced the title.

Elsewhere

The foundational narrative of landscape architecture is certainly a transatlantic affair, and Olmsted's trip to Birkenhead is often celebrated as the moment of inception. But similar histories can be traced in other countries: a few examples from Europe will reveal the way in which landscape architecture has emerged from earlier traditions of garden and park design, though they will also demonstrate the way that different histories and cultural characteristics have shaped the developing character of the discipline in each nation. In France, where the English style of gardening had been widely adopted in the 18th and 19th centuries, the engineer Jean-Charles Adolphe Alphand (1817–91), supported by the horticulturist Jean-Pierre Barillet-Deschamps, constructed a number of public parks in connection with Baron Haussman's remodelling of Paris. The most striking of these was the Parc Buttes-Chaumont in the north-east of the city, which includes the site of a former limestone quarry whose towering ivy-clad cliffs are topped by a replica of the Roman Temple of Vesta. The French word for 'landscape architect' is *paysagiste* (a very approximate translation would be, 'countryside-ist'), but the profession was only officially recognized after the Second World War, when the first training courses were established at the horticultural school at Versailles. In the latter decades of the 20th century, re-emergent French traditions of formality, given new life by a chaotic post-modern collaging of ideas, became very influential, as they offered a bracing alternative to tired picturesque scene-making.

In Germany, the most significant figure in the transition from landscape gardening to landscape architecture was Peter Joseph Lenné (1789–1866), gardener to the king of Prussia. In addition to his royal commissions, he laid out some of Germany's earliest public parks, including the Friedrich-Wilhelm Park in Magdeburg, the Lennépark in Frankfurt (Oder), and the Tiergarten and Volkspark Friedrichshain in Berlin. It was not until 1913 that the Bund Deutscher Gartenarchitekten (Association of German Garden Architects) was founded in Frankfurt am Main, only changing its name to the Bund Deutscher Landschaftsarchitekten (Association of German Landscape Architects) in 1972. The design of communal green open space received attention during the Weimar Republic, but the development of landscape architecture was compromised by the involvement of prominent practitioners with National Socialism. During this period, landscape architects not only worked on planting alongside the newly built motorways, but also, notoriously, on the 'Germanization' of rural landscapes in the conquered east. After the Second World War, landscape architecture was quickly re-organized, at least in the west, and practitioners played a significant role in the reconstruction of the war-damaged country. A series of biennial Bundesgartenschauen (Federal Garden Shows), the first of which was held in Hanover in 1951, demonstrated the way in which landscape architecture could transform derelict and war-damaged sites into permanent parkland.

The Dutch park designer Jan David Zocher Jr (1791–1870) was influenced by Brown and Repton, and his Vondelpark in Amsterdam, first opened in 1865, is a romantic, naturalistic park in the English manner. However, the Netherlands also has a history of winning land from the sea and, in the 20th century, comprehensive landscape planning was required to create completely new settlements and landscapes on the polders. The botanist Jacobus Pieter Thijsse (1865–1945), often regarded as the father of the Dutch ecological movement, contributed an internationally influential idea: he suggested that every town or district should have an 'instructive garden', where people could learn about the nature on their doorsteps. Thijsse was concerned about the loss of species in the countryside, through such practices as the draining of swamps and the forestation of heathland.

The Netherlands is highly urbanized and the countryside is very obviously a human creation, yet there is a yearning for contact with nature. Perhaps as a result, the country produces some of the most interesting new ideas in landscape architecture and urbanism. The title of 'landscape architect' has been legally protected there since 1987.

Landscape architecture is now a global discipline, but in many countries it is still in its infancy. Over 70 national associations are affiliated to the International Federation of Landscape Architects, a list that begins alphabetically with Argentina and Australia and ends with Uruguay and Venezuela. The list includes countries as populous as the United States, China, and India and as small as Latvia and Luxembourg. The number of practitioners varies widely too. The Canadian Society of Landscape Architects has over 1,800 members, the Fédération Française du Paysage has over 500 members (but only represents one-third of practitioners), and the Bund Deutscher Landschaftsarchitekten has about 800 members. The Landscape Institute (UK) has over 6,200 members, while the Irish Landscape Institute, only founded in 1992, has 160. By far the largest association is the American Society of Landscape Architects with around 15,500 members. Although there are efforts to standardize education, qualifications, and the requirements for registration, these still vary significantly from country to country, depending upon local institutional structures and laws. Even in the matter of education there is considerable diversity. In some countries, landscape architecture is taught in association with horticulture, agriculture, or gardening. In others, it is the bedfellow of architecture, planning, and urban design. Elsewhere, it may be found in a school of forestry or environmental sciences. While the similarities between the landscape architecture programmes in these different sorts of institution will greatly outweigh the differences, there is no doubt that each will have its distinctive emphasis or flavour.

After reading this potted history, I hope you will have gained some sense of the scope of landscape architecture, but you might now be wondering if the discipline has any definable core. The next chapter will consider the range of activities that come under the general umbrella of landscape architecture and look at various attempts to define the essence of the discipline.

Chapter 2
The scope of landscape architecture

It is time to look at what landscape architects actually do. This small selection of case studies is an attempt to convey the diversity of contemporary practice and to give you a feeling for the broad scope of the discipline. The chapter describes four projects chosen to exemplify: high-profile master-planning; visual impact assessment; art-inflected urban design; community engagement. They range from the worthy to the flamboyant. You might like to consider whether they all fall under the rubric of 'improvement and place-making'.

Gardens by the Bay, Singapore (2006–ongoing)

The first project is the result of a design competition organized by the National Parks Board of Singapore, which was looking for a design team to master-plan their Gardens by the Bay project (Figure 3), a horticulturally themed attraction in the new downtown area of Marina Bay, constructed on a reclaimed waterfront. Ultimately there will be over 100 hectares of tropical gardens comprising three distinct gardens—Bay South, Bay East, and Bay Central. The commission for the first phase of this huge project—Bay South—was awarded to a British design team led by landscape architects Grant Associates in conjunction with architects Wilkinson Eyre. Grant Associates' master-plan was inspired by the shapes of the orchid, Singapore's national flower,

3. The commission for the first phase of Singapore's Gardens by the Bay project was awarded to a British design team led by landscape architects Grant Associates in conjunction with architects Wilkinson Eyre

and the project enjoyed political support at the highest level. The plan weaves together nature and technology, incorporating two artificial biomes, the Flower Dome and the Cloud Forest Dome, designed by the architects to house plants from Mediterranean and Tropical Montane climates, respectively. The landscape architects designed strikingly impressive Supertrees, some as tall as 50 metres, which form part of the cooling system for the conservatories, but also carry towering displays of epiphytic plants, ferns, and flowering climbers, and are illuminated at night. The technologies concealed in the Supertrees mimic the ecological functions of real trees; they include photovoltaic cells which power some of the lighting and they collect and channel rainwater for use in irrigation and water displays.

The Gardens by the Bay have been described as a horticultural Disneyland and a scene from *Alice in Wonderland*, but also as a triumph of environmentally conscious design. As intended, they have captured the attention of the world's media. Projects of this scale and ambition are rare, even for the most renowned design offices, but the Gardens nevertheless exemplify many features of contemporary practice. For instance, the project involved an interdisciplinary team of designers, not just landscape architects and architects, but also specialist environmental design consultants, structural engineers, visitor centre designers, and communication specialists. The location and condition of the site, reclaimed land on a waterfront, is also characteristic of many large projects over recent decades, such as the Havneparken (Harbour Park), Copenhagen, Denmark (completed in 2000), the Daniaparken, Malmö, Sweden (completed in 2001), and Shanghai Houtan Park, Shanghai, China (completed in 2010).

Hirddywel Wind Farm landscape and visual impact assessment, Wales (2010)

While the Gardens by the Bay were designed to be conspicuous and totemic, this project, which belongs in the realm of strategic landscape planning rather than master-planning or site design, will be judged to have been successful if the public remain unaware that it ever occurred. Landscape architects have often been involved in minimizing intrusions into the countryside. They are often involved in applications to extend quarries, for example, or to open-cast for coal. In Britain in recent years the siting of wind turbines has been one of the most hotly contested land-use planning issues. Whatever the merits or demerits of particular turbine designs, these machines seem to bring out an almost visceral loathing in some rural communities, perhaps because they are regarded as alien impositions which only benefit distant cities. Though people can be generally in favour of an idea, clean energy or high speed transportation, for instance, they come out in opposition if it leads to proposals on their doorstep, a phenomenon which goes by the acronym NIMBYism (not-in-my-backyard). The height and the number of turbines in a proposal have a bearing upon its acceptability, as does the existing topography. Landscape architects have become experts in modelling and mapping the 'zones of visual intrusion' for wind turbine proposals, and indeed for any large addition to the landscape. Using photomontage techniques and computer visualizations they can show what any proposal will look like from a variety of key viewpoints. They are often also involved in the design of mitigation proposals, which may include screening earthworks or planting, to reduce such impacts.

From a technical stance, high and open areas of land are particularly suitable for the siting of wind-farms, but such areas are often highly valued for their existing landscape character. The Welsh government is committed to doubling the amount of energy generated from renewable sources by 2025 and has identified seven strategic research areas where they believe large scale wind farms could be developed. AMEC Environment and Infrastructure UK Ltd, a company which provides a range of services, including landscape architecture, was commissioned by NUON Renewables to carry out a landscape and visual assessment of its proposal to build the Hirddywel Wind Farm in Powys. This assessment was complex, because the company was also proposing another wind farm to the east and already operated another facility nearby which it hoped to reconfigure with fewer but taller turbines. The consultancy's visualizations made it possible to assess the cumulative impacts of the proposals and led to a reduction in the number of proposed turbines from 13 to 9. The assessment was completed in 2010 but at the time of writing the proposal is still in the planning process.

The Hirddywel Wind Farm is a good example of a major infrastructure proposal which could have great benefits at the national scale, but is also likely to have significant local impacts. In countries with well-developed planning systems, getting approval for such developments can be a long and involved process. Landscape architects can assist at all stages, from pre-application assessment, through the planning process itself, which may involve presenting evidence to a public inquiry, to the design and implementation of mitigation measures. It is also worth pointing out that while private developers may employ landscape architects, in many countries local authorities also do so. Like the rival forensic pathologists called by the prosecution and defence in many a courtroom drama, landscape architects sometimes find themselves on opposite sides in particularly contentious planning inquiries.

Les Boules Roses, Montreal, Canada (2011)

The Canadian landscape architect Claude Cormier (1960–) has made his name with witty and artistic interventions in urban life. Many of these, such as his Sugar Beach in Toronto or Clock Tower Beach on the Quai de l'Horloge in Montréal's Old Port, are intended to be enduring additions to the urban fabric, but he is also known for his temporary installations and Les Boules Roses is one of these. Strung across Sainte-Catherine Street East in Montreal's gay village were 170,000 pink resin balls, helping to transform a workaday street into an enchanted promenade for the Aires Libres festival in 2011. The balls came in three different sizes and five subtle shades of pink. They were laced across the street to form a canopy which intermingled with the branches of existing avenue trees, casting a dappled shade and stretching for 1.2 kilometres between Berri and Papineau Streets. The installation was set out in nine sections, with variegated patterns to create a range of moods along the route.

Cormier is the protégé of the American landscape architect Martha Schwartz (1950–)—at one time the iconoclastic *enfant terrible* of the discipline, though now one of its most respected educators and practitioners. Coming into the discipline from a background in art, Schwartz's early work, which incorporated unconventional materials such as plastic trees and flowers, Plexiglas chippings, and even, infamously, shellacked bagels, was often a deliberate provocation, inviting rejection by the landscape architecture world: 'Can this really be landscape architecture, or is it something else?'. Practitioners like Schwartz, Cormier, and the German office Topotek 1 are the ludic wing of landscape architecture. They enjoy overturning assumptions and confounding expectations. Yet alongside the playfulness there has to be an understanding of site, context, and the needs of users, particularly when the design intervention is made to last. The best of such practice recognizes all of this, so that the projects are not only fun but also functional.

West Philadelphia Landscape Project (commenced 1987)

The West Philadelphia Landscape Project was an action research programme, initiated by Anne Whiston Spirn, who is a landscape architect, educationist, author, photographer, and activist. It was originally based in the Department of Landscape Architecture and Regional Planning at the University of Pennsylvania where Spirn was professor from 1986 until 2000 when she moved to Massachusetts Institute of Technology, Boston. From the outset the programme sought to integrate research with teaching and community service. In particular it involved the design and construction of a series of community gardens in the socially deprived neighbourhoods of West Philadelphia. These were small, incremental improvements which could not hope to solve all the problems of poor housing stock, inadequate infrastructure, poverty, and unemployment, yet in addition to brightening up the urban landscape they served as catalysts for other forms of community development. After 1995, the project produced an offshoot, the Mill Creek Project, which was a collaboration between students and researchers at the University of Pennsylvania and teachers and pupils at the Sulzberger Middle School in West Philadelphia. It was organized around the creation of a new middle-school curriculum entitled 'The Urban Watershed'. It sought to raise environmental awareness of the school's locality and was centred on the presence of a culverted stream, the Mill Creek, which had once run through the field upon which the school playground had been built. The presence of this buried watercourse had caused numerous problems of flooding, subsidence, and outright collapse. The landscape architects involved in the project were able to draw attention to the difficulties which ensue when urban development takes place upon a flood plain. They were also able to suggest ways in which undeveloped land might be redesigned to detain storm water, thus reducing the risk of flooding while providing socially valuable open space.

I cannot give an end date for the West Philadelphia Landscape Project. The timeline on its website runs out in 2009, but it has a blog which still carries the occasional post and it seems likely that the influence of the project upon these troubled neighbourhoods and their residents will play out over generations, while the gardens created are a tangible legacy. By any measure, the West Philadelphia Landscape Project must count as one of the longest-running community engagement projects in the field of landscape architecture, and certainly the most recognized and celebrated. In 2001, it was cited as a 'Model of Best Practice' at a White House summit for 40 leading scholars and artists in public life. In 2004, it won the Community Service Award from the American Society of Landscape Architects. Spirn's current book project, *Top-Down/Bottom-Up: Rebuilding the Landscape of Community*, is based upon her experiences of the project over 25 years.

Is there a core?

Quite deliberately, I selected four projects for this chapter which seemed to have little in common, except that landscape architects played the leading role in each. Perhaps, on further reflection, we might start to find commonalities. The designers of the Gardens on the Bay and Les Boules Roses were each trying to create a visual spectacle and a sense of festival which would appeal to visitors. The West Philadelphia Landscape Project and the wind farm study in Wales were both concerned with the consequences of siting development and infrastructure. Nevertheless, it would be easy to find another batch of landscape architectural projects, apparently as diverse as these four, and then to find four more. The variety of projects is mirrored by a diversity of approaches. Some landscape architects take pride in the invisibility of their work. When mitigating the impact of a proposed motorway or power transmission line they want their work to blend into the surrounding landscape as harmoniously as possible. Others strive for effects that are startling, amusing, or theatrical and would be dismayed if they thought their artistry was being overlooked. Some place great importance upon working in collaboration with communities, sublimating any egotistical urges to which they might be prone in favour of a socially sustainable outcome. Others cannot abide compromise and feel that the best design work expresses a singular vision. Still others may combine attributes within a practice, or shift their approach depending upon the site, client, or brief.

If there is such diversity of opinion and if the sorts of projects landscape architects work on are so varied, is it possible to come up with a definition of the discipline or to say what is essential to it? The call for definitions and boundaries, I would argue, is often a symptom of insecurity. The professionalization of a discipline involves setting standards for entry controlled by examination and with this there is an urge to determine what should be the core knowledge and skills that practitioners should possess. The positive side of this is that it offers protection to clients and the public; professionals should know what they are doing. The case is easy to make for medicine, where no one would trust an unlicensed brain surgeon, and for civil engineering where building a bridge without carrying out the necessary calculations has obvious implications for public safety, but it is not so easy where the skills involved may be widely available in the community and where some possible harms are diffuse and may not be identifiable in the short term. The fact that architects, urban planners, and landscape architects engage, to various degrees, in public consultation and participatory engagement shows that lay knowledge and opinions are valued. One might even say that the discipline is based on professionalized lay knowledge. Indeed, some of the longer term harms that manifest in badly planned or designed housing areas or new towns are often blamed upon a failure to understand what people actually want or need from such developments. The downside of professionalization, then, is that it can lead to a protectionist attitude or an exclusionary 'closed shop' which keeps clients, users, and members of kindred professions outside. Much of the talk about standards, codes of conduct, and accreditation is an attempt to control particular areas of work and is thus commercially motivated and often suspect.

One of the manifestations of professionalization is the urge to define a core curriculum, but in a discipline as diverse and wide-ranging as landscape architecture, this is becoming an impossible demand. Landscape architecture may have a fluid core but it does not have a fixed essence. It has borders with other disciplines, including engineering, art, architecture, urban planning, and urban design, but these are not fixed boundaries and they are permeable. Nevertheless, it remains its own discipline and cannot be assimilated by its neighbours. A useful way to conceptualize this is to think of landscape architecture as an extended family. In this family, there will be people who do the same sorts of things that Frederick Law Olmsted and Calvert Vaux once did: they will design parks or systems of parks, although they may not necessarily share Olmsted's views about the appropriateness of pastoral scenery in the midst of the city. There will be others who have never designed a park, although they have worked with engineers designing transport infrastructure. Others have specialized in the domestic market and work almost entirely on gardens. Some have spent their careers working alongside foresters, helping them to plan their plantations and their operations in ways that are visually and environmentally harmonious. Others still are happiest when working in the city, designing or refurbishing urban squares and pedestrianized streets. Pick two individuals from this range and you might find that their working lives are so different that it is difficult to conceive of them as members of the same profession, but between them, and linking them to the rest, are webs of resemblance. Landscape architecture's openness is perhaps its greatest strength and its permeable boundaries should be a model for other disciplines.

Attempts to define the discipline usually fail (I include the IFLA definition quoted at the end of the Preface) and I would argue that this is inevitable. Most of them are prolix and wordy, trying to capture all of the assorted activities in which landscape architects are engaged. The late Marlene Hauxner, Professor of Landscape Architecture at the University of Copenhagen, wrote a book entitled *Open to the Sky* and I have heard it said that landscape architecture is concerned with the planning and design of everywhere which does not have a roof, but even this promising definition stumbles, because whole books have been written about *interior* landscape design. This is the problem with essentialist definitions; some counter-example can usually be found which just does not fit. I also like the assertion made by Tom Turner, who teaches at the University of Greenwich, London, that landscape architecture is about 'making good places'. He emphasized the word 'good' to stress that making any old place would not do—but of course his exhortation is very general and leaves open the question of what is to count as good. The answer to this question could involve ecology, psychology, sociology, politics, aesthetics, and more besides. These are some of the things which place-making and improvement seem to involve in the 21st century.

Chapter 3
Modernism

By the time of the 1899 meeting in New York which formalized the landscape architecture profession, a rejection of tradition was sweeping through the entire world of the arts. 'Modernism' meant different things to different disciplines, though a pervasive concern was the search for forms of expression relevant to the new social conditions ushered in by industrialization and technological advancement. Modernist thinking in fine art (particularly painting) and in architecture had very different trajectories, but both had a powerful influence upon the nascent discipline of landscape architecture.

The influence of Modern Art

Sir Geoffrey Jellicoe (1900–96), doyen of 20th-century British landscape architects, theorized that landscape design bore a specific relation to the visual arts, particularly painting. A landscape design, he argued, takes a long time to create, for even if the design phase can be done quickly, the construction, which might involve moving great quantities of earth, puddling the clay linings of large lakes, or planting hundreds of trees, is likely to be time-consuming and generally far beyond the powers of any individual working alone. Even when the landform has been created and the planting has gone in, it can take many seasons of growth before the landscape begins to resemble its intended form. These constraints make experimentation difficult. The painter, on the other hand, is in a relatively enviable situation. The material requirements are fewer: a studio, an easel, some canvasses, and some paint. Painters, Jellicoe argued, are thus able to serve as aesthetic pathfinders, while the best a landscape architect can do is to keep up with them. In the 18th century the designers of landscapes had paid close attention to works of art, drawing their ideas from paintings by such luminaries as Nicolas Poussin (1594–1665), Claude Lorrain (*c.*1604–82), and Salvator Rosa (1615–73). However, in the 19th century, he claimed, things had gone seriously wrong. The connection with painting had been severed by an excess of enthusiasm for horticulture, boosted by the stream of new species and varieties sent back to Britain by adventurous plant-hunters, and by technological advances like the steam-heated glass-house, which encouraged a competitive attitude towards horticultural display. In the 20th century, however, landscape design rediscovered its association with art—but art, in the meantime, had moved on.

Even in the days when landscape was still a fashionable genre of painting, the art world did not generally set great store by the accurate depiction of topography. As the geographer and art historian Peter Howard has observed, if you are looking for an accurate record of a landscape, you are more likely to find it in the work of a less celebrated artist than a famous one. Indifference to verisimilitude was such that Henry Fuseli in his role as secretary of the Royal Academy would exclude works which were committed to the 'tame depiction of a given spot'. In any case, the role of recording landscape passed on to photographers who could do it more accurately and quickly. Artists of serious purpose reacted to the arrival of photography by turning to abstraction. As Howard has noted, there are Cubist landscapes, Surrealist landscapes, and Expressionist landscapes, but the places depicted (if this word can even be used) are of much less consequence than the theory and the method being explored. Nevertheless, this was the art—abstract art—which Jellicoe believed could show landscape designers the way forward.

There were indeed landscape designs which drew direct inspiration from abstract painting. The architect Gabriel Guévrékian (*c*.1900–70) designed a Garden of Water and Light for the Exposition internationale des Arts Décoratifs et industriels modernes, in 1925, now remembered as a showcase for Art Deco, and this persuaded Charles de Noailles to commission a triangular abstract garden for his villa at Hyères. Art Deco was influenced by Cubism and by a fascination with technology, but its lack of concern for function set it apart from other currents in architectural Modernism. Guévrékian's, angular gardens with their use of concrete, geometrical patterns and sparse planting must have perplexed many gardeners, but designers saw them as a categorical break with both horticultural and naturalistic traditions. These were, however, gardens to be principally admired for their style, not 'rooms outside' to be used. Nevertheless, Fletcher Steele (1885–1971), an American landscape architect, was sufficiently impressed by Guévrékian's work to write an article with the title 'New Pioneering in Garden Design', a Modernist call to arms which initially went unheeded by his peers. Steele began to experiment with Modern ideas in his own design practice, hitherto based upon Italianate formality or the English Landscape style. At Naumkeag, Stockbridge, Massachusetts (1925–38) he took a Renaissance idea, a series of flights of steps rising through woodland, and produced a simplified and much photographed Modernist version, the Blue Stairs, with elegantly sweeping white metal handrails.

Another practitioner influenced by the trend towards abstraction was the Brazilian polymath Roberto Burle Marx (1909–94), a painter, sculptor, jewellery designer, and creator of theatrical sets as well as a botanist, plantsman, and landscape architect. He thought of himself primarily as a painter and his colourful canvasses resemble those of Arp and Miró. Their biomorphic shapes and brilliant colours are also found in his planting plans. By using single-variety blocks he could metaphorically paint landscapes with foliage, as he demonstrated in his celebrated garden for the Monteiro family on their estate near Petrópolis (1946). Some of his work is highly patterned, such as the pavement he designed for the three-mile long promenade of Copacabana Beach in his native Rio de Janeiro (1970). In addition to an impressive catalogue of private commissions, he worked on several notable public projects, including a roof garden for Oscar Niemeyer's Ministry of Education (1937–45) in Rio de Janeiro, and the planning of Brasilia (1956–60) with the architect Lucio Costa.

Another important transitional figure was Thomas Church (1902–78) who trained at Berkeley and Harvard, then travelled on a scholarship to Spain and Italy where he realized how similar the climate in California was to that of the Mediterranean and how conducive it could be for outdoor living (Figure 4). During the Depression he opened a small office in San Francisco and gradually built a career designing gardens for affluent middle-class clients rather than the ostentatiously rich. Fletcher Steele's embrace of Modernism made it easier for Church to throw off the yoke of symmetry. His relaxing gardens with their timber decking and free-form pools became a recognizable component of the West

4. Plan of Thomas Church's Kirkham Garden (1948): the garden as an outdoor room for living

5. Thomas Church's Donnell Garden, Sonoma County, California (designed with Lawrence Halprin, 1954) became emblematic of the West Coast lifestyle

Coast lifestyle, and his way of designing soon became known, inevitably, as the Californian Style. He promoted it through articles in lifestyle magazines, the most influential of which was *Sunset*, a publication aimed at those settling in California from the East. Church was influenced by abstract art: like Burle Marx, he seems to borrow shapes from Arp and a characteristic ploy was to play off piano curves against grids of paving or zigzagging timber benches. Though Church was influenced by Cubism and Surrealism, it was his meeting with the Modernist architect Alvar Aalto in Finland in 1937 which drove the development of his mature style. Respecting the climate, his gardens made little use of lawn which in the West Coast climate would need constant irrigation. Instead Church used paving, gravel, sand, redwood decking, and drought-tolerant groundcover planting (Figure 5). He designed some 2,000 gardens but his masterpiece is generally thought to be the garden of the Donnell Residence, Sonoma County, California (designed with Lawrence Halprin, 1954) where many of these elements were brought together in harmonious perfection. Church's decision to build the extensive decking around the existing live oaks on the site is also celebrated, while Adeline Kent's lissom sculpture for the pool became emblematic of the sybaritic Pacific Coast way of life.

The influence of architectural theory

Some influential landscape architects have also been qualified architects or have worked closely with them. The two disciplines have a long affinity, so developments in architectural theory inevitably impact upon landscape architecture. Modernism in architecture, according to the historian Nikolaus Pevsner, emerged from Art Nouveau's refusal to be hamstrung by the past, and the English Arts and Crafts Movement's demand for excellence and integrity in design. When these trends were united with the enormous potential of industrial technology and with new materials like steel and glass, the way was open for a break with all traditions. More radical than its antecedents, architectural Modernism turned against both individual craftsmanship and extraneous decoration in favour of a pure doctrine of functionalism. A few quotations from its prophets and high priests will indicate its tenor. Adolf Loos (1870–1933), for instance, declared that 'all ornament is excrement', while Le Corbusier (1887–1966) believed that the house should be 'a machine for living in'. Ludwig Mies van der Rohe (1886–1969), director of the Bauhaus design school from 1930–3, left us the pithy minimalist dictum that 'less is more', as well as the maxim that 'God is in the details'. Modern architecture, true to the spirit of its age, was to be a stripped-down, functional creation, whose aesthetic interest came not from applied ornament but from its apparent fitness for purpose and honesty in its use of materials. Industrial mass production and prefabrication would make excellent design available to large swathes of humanity, thus Modernism was often coupled to a progressive social vision. Le Corbusier's career is instructive. There is no doubt that he was one of the century's creative geniuses, and his smaller projects, such as the Villa

Savoye (1928–31) or the chapel at Ronchamp (1950–4), are rightly recognized as 20th-century masterpieces. However, like many self-confident architects who have turned to city planning, his prescriptions for new urban form could have disastrous consequences. He suggested tearing down large portions of central Paris in order to implement the Plan Voisin (1925), replacing the higgledy-piggledy richness of diverse old quarters with an unlovely grid of tower blocks, each identically cruciform in plan, and unrelentingly imposed across the face of the city. Mercifully this was never implemented, but lesser architects and planners picked up on such sanitizing ideas of grand reconstruction with consequences now recognized as dire. The demolition in 1972 of the Pruitt-Igoe housing project in St Louis, Missouri, built only 16 years earlier on rational Modernist precepts but notorious for its mounting catalogue of social ills, is often said to have been a turning point. It was, claims the architect and critic Charles Jencks, the end of the Modernist dream.

It is not surprising that landscape architects were swept up in the momentum of Modern architecture's heady charge, even though they struggled to apply the doctrine of functionalism or find uses for concrete, steel, and glass. One of the first to write enthusiastically about Modernism was Christopher Tunnard (1910–79), a Canadian-born designer who settled in England in 1928. His book, *Gardens in the Modern Landscape*, was the first manifesto for a Modern landscape architecture. Tunnard not only ridiculed Victorian designers for their fussy ornaments and elaborate herbaceous borders, but even turned on the great Corbusier for illustrating so many of his buildings in a pastoral setting. For Tunnard, the landscape had to be designed on the same rational, purposeful principles as the buildings. In addition to admiring Modernist architecture, he was also enthusiastic about traditional Japanese architecture and garden design, which, he thought, came to the beautiful by way of the functional. He designed two notable Modern gardens in England, one for his own house, St Ann's Hill, Chertsey, Surrey, designed by Raymond

McGrath, the other for the home designed by the Russian émigré Serge Chermayeff at Bently Wood, Halland, Sussex (both 1936–7), but he found Britain resistant to the new thinking and so, when invited to teach at Harvard Graduate School of Design by Walter Gropius (1883–1969), founder of the Bauhaus but by this time an émigré, Tunnard left for America. Eventually he taught at Yale University, where his interest shifted from design towards urban planning and historic preservation. Indeed, as early as 1946, he began to repudiate the dogmas of Modernism, warning that 'There is a dangerous fallacy in thinking that a certain kind of architecture or planning is intrinsically "better" than another.'

The Harvard Rebels

Harvard, where Tunnard initially went to teach, has a close connection with landscape architecture and with the 20th-century shift toward Modernism. The subject had been taught there since 1900 when a course had been established in memory of Charles Eliot, the son of the president of the university. In 1893, Eliot had become a partner in the practice of Olmsted, Olmsted and Eliot. His partners were Frederick Law Olmsted and his nephew and stepson John Charles Olmsted (1852–1920), but the elder Olmsted's health was soon failing and Eliot found himself leading the firm, which was the officially appointed landscape architect to Boston's Metropolitan Park Commission. He had a difficult time trying to persuade the commissioners to produce a comprehensive plan for the city's park system and his mounting frustration might have contributed to his untimely death from meningitis in 1897 at the young age of 37. The programme set up by his father was headed by another Olmsted, Frederick Law Olmsted Jr, son of the designer of Central Park. The links between the university and the emerging discipline could hardly have been stronger, but by mid-century landscape architecture teaching had got into a rut. The emphasis remained upon Olmstedian visions of pastoral landscape, with a general assumption that naturalistic design was inherently superior to anything formal or obviously man-made. However, when Gropius arrived in 1937 and the decision was taken to merge the three departments of Architecture, Landscape Architecture, and City and Regional Planning as the Graduate School of Design, all this was set to change. Landscape architecture and architecture students would collaborate on studio projects and in this way Bauhaus thinking began to infiltrate the landscape architecture curriculum. Tunnard had been brought in expressly for his avant garde ideas and was to be a catalyst for change. Three mature students also took up the Modernist cause. Their names were Garrett Eckbo (1910–2000), Dan Kiley (1912–2004), and James Rose (1913–91). Collectively they are often referred to as the Harvard Rebels.

Rose's career, principally as a garden designer, has been somewhat overshadowed by those of his illustrious contemporaries, although there is now a study centre located in his former home in Ridgewood, New Jersey. In the 1930s, his articles for the magazine *Pencil Points* attacking both axial and picturesque approaches to design were provocative and influential. Dan Kiley was so disenchanted with Harvard's conservative outlook that he left without graduating, but after the War, through his friendship with the architect Eero Saarinen, he was taken on to work on the Palace of Justice at Nuremburg and while in Europe was able to visit many historic formal gardens where he acquired the design vocabulary of the *alleé*, *bosquet*, and *boulevard*. Later he worked with Saarinen on the J. Irwin Miller House in Columbus, Indiana (1957), where he was able to use some of these elements in a Modern idiom. Kiley recognized (as too did Jellicoe) that the Modernist garden and the Classical garden were not, after all, so far apart in spirit and could be successfully fused. Kiley took from Modernism its stripped-down aesthetic with its clean lines and crisp geometries, but he was happy enough to combine it with a Classical symmetry where it seemed appropriate. Thus his design for the Henry Moore Sculpture Garden at the Nelson Atkins Museum of Art, Kansas City (1987–9), with its gentle terraces and clipped hedges pays homage to Le Nôtre as, in its individual way, does his design for the plaza surrounding I. M. Pei's Allied Bank Tower, Dallas, Texas (1986), which incorporates 263 bubbler fountains and a grid of cypress trees in circular granite planters.

Of the three Rebels, it was Eckbo who pursued the social vision of Modernism with the most vigour. In 1939, he returned to California where he worked initially for the Farm Security Administration, helping to design settlements for migrant workers, refugees from the Dust Bowl of Oklahoma and Arkansas. He was able to bring some of his Bauhaus training to bear on the design of housing that would not only meet basic needs, but also foster a cheerful sense of community. It was a philosophy that would serve him well after the Second World War when there was a great demand for new homes. His desire to create good environments for the working classes is often contrasted with Church's work for more prosperous clients. Forming a partnership with Robert Royston and Edward Williams in 1945, the firm initially competed for garden work against Church, but Eckbo had a broader vision of what landscape architecture might do and what it could become. The office took on bigger projects, for campuses, parkways, urban squares, and the surroundings of industrial buildings and power stations. Notable projects included Downtown Mall, Fresno (1965), Union Bank Square, Los Angeles (1964–8), and the open spaces at the University of New Mexico, Albuquerque (1962–78). Eckbo became chair of the Berkeley Department of Landscape Architecture in 1963 and in 1964 the practice known as EDAW (Eckbo Dean Austin and Williams) was formed, which went on to become one of the most influential landscape and urban design firms the world has ever known. Eckbo was surrounded by good designers who shared his ethos, but he was the one who could best articulate their mission, which he did most effectively in his *Design for Living*, a book published in 1950 which now has the status of a classic. It was an attempt to meld the highest theory with the most down-to-earth practicalities and to show that aesthetic and social objectives could be combined. He was dismayed by the fragmentation and dysfunctionality he found in the urban environment, but thought it was the role of the planner or designer to work with people's best cooperative instincts. Lawrence Halprin (1916–2009), who trained at Cornell Agricultural School before going to Harvard in 1940, just missed being a classmate of Rose, Kiley, and Eckbo, but he was much influenced by Gropius, Tunnard, and also Marcel Breuer (1902–81), another former teacher from the Bauhaus who had fled to the United States. After wartime service, Halprin went to San Francisco where he worked for Church. Though he got on well with his employer, he found private garden work restrictive. In his own words, he wanted to break out of the 'garden box' and work on 'broader scale community work', so after four years he left to set up his own practice. Among his many contributions to the development of the discipline, he is probably best known for his striking urban parks, particularly Lovejoy Plaza (1966) and Ira Keller Park, both in Portland, Oregon, and both abstracting in concrete the sort of mountain scenery in which Halprin loved to walk. Freeway Park, in Seattle, Washington (1970), which bridged the chasm of a freeway cutting, is also celebrated, as is his pioneering work in urban regeneration at Ghirardelli Square, San Francisco. In the 1960s, he helped to plan a new community on the Californian coast known as Sea Ranch which was driven by an ethos of environmental respect and 'living lightly on the land'. Halprin was an innovator who sought new collaborations and invented new creative methods. With his wife Anna, a dancer and choreographer, he developed a 'motation' to record movements through an environment, and made connections between creating a design and writing a score. He sought, in the spirit of the Bauhaus, to bring together collaborative teams combining insights from different disciplines and was an early advocate of involving citizens in the design process.

Modernism elsewhere

The United States was not the only country to produce a group of Modernist rebels. In Denmark, G. N. Brandt (1878–1945) turned against Beaux Art historicism, but was influenced by spatial clarity of English Arts and Crafts gardens. He is particularly remembered for the geometrically ordered Mariebjerg Cemetery he designed at Gentofte (1925–36). Many landscape architects served an apprenticeship in Brandt's studio, the most illustrious of whom was C. Th. Sørensen (1893–1979), whose most significant contribution to design was to bring elements of the Danish cultural landscape into his architectonic practice, as he did with the use of hedges to form elliptical enclosures, first in his allotment gardens at Naerum (1948) and then in the sculpture garden associated with the Angli IV factory in Herning (1956). As in America, a typical attitude among Danish Modernists was to regard gardens and landscapes as adjuncts to architecture, outdoor rooms which formed part of the overall composition. Modernism was strong in Sweden too, where it was coupled with progressive social ideals. Holger Blom (1906–96) was an urban planner not a landscape architect, but in 1938–71 he was Director of Parks in Stockholm and promoted the enlightened policy that parks should be seen as social necessities—as significant for civilized life as houses with hot and cold running water. They needed to be planned for active use and they needed to permeate the whole of the city. The landscape architect who helped Blom realize this goal was Erik Glemme (1905–59). While America had produced the California School with an emphasis on private gardens, socially democratic Sweden produced the Stockholm School of Park Design, dedicated to public service. The Stockholm School avoided formal and picturesque idioms, but found inspiration in the regional landscape. Though naturalistic in its visual style, it embraced rational planning, functional goals, and modern materials.

Modernism in architecture was iconoclastic, but in seeking to overthrow historical styles and rigid formulas, it ultimately created straightjacket rules of its own. The International Style was promoted in the 1930s as the only style appropriate for the age, and it was one which, as its name implied, could be applied anywhere in the world, regardless of history, culture, or climate.

The steel-framed skyscraper with curtain glass walls became the favoured style of international finance, and the business districts of cities as far flung as Bangkok, Toronto, Melbourne, and Singapore came to look like clones of Manhattan. Landscape architecture was saved from this homogeneous fate in part by the 18th-century injunction to 'consult the genius of the place', but also by intractable regional variations in climate, soils, and vegetation. Modernism had never truly replaced such staple landscape materials as earth, water, and plants. Many of the best ideas from the Modern period have survived. Care about materials, an emphasis upon space, a rational approach to site planning, an aesthetic delight in efficient and elegant detailing—these are all part of the positive legacy of Modernism. Above all, there remains the notion that landscapes should be functional, something which will be explored further in the next chapter.

Chapter 4
Use and beauty

Since John Dixon Hunt first suggested the notion of the 'three natures', there has been much speculation about whether the number of categories needs to be expanded. If a farmed landscape is deliberately left unmanaged as part of a policy of 're-wilding', is the ensuing landscape part of first nature (wilderness), second nature (cultivation), or third nature (designed with aesthetic intent)? The very term 'landscape' carries with it the notion of a hybrid between nature and culture. Some say that we need the concept of 'fourth nature' to cover such conceptually complex places as managed nature reserves, reclaimed landscapes, restored habitats, and so on. But even without the complication of fourth nature, there are problems enough about making aesthetic intent the criterion for deciding between the commonplace landscapes of second nature and the pleasure grounds of third nature. Aesthetics is often an issue, even for workaday places.

This is easily illustrated by the case of the *ferme ornée* (literally the ornamented farm), an idea that caught on among 18th-century English landowners. We owe the term to Stephen Switzer (1682–1745) an early exponent of the English Landscape School. It referred to an estate laid out partly according to aesthetic principles and partly for efficient farming. The most famous example was the poet William Shenstone's garden at the

Leasowes, Shropshire, which was visited by many eminent figures including William Gilpin, Thomas Gray, Oliver Goldsmith, Samuel Johnson, and Thomas Jefferson. If a productive farm could be laid out as a pleasure ground, why then could not other sorts of useful places, such as forests, cemeteries, or reservoirs? Even if a place is essentially utilitarian, why should it not also be pleasing, or at least not be ugly?

Even without much 'improvement', everyday landscapes can give aesthetic pleasure. One only has to think how often farmland has been the subject of painting; Breughel, Hobbema, Van Gogh, and Constable all found sufficient interest in fields to want to paint them. Picturesque painters were not averse to moving things around on the canvas or exaggerating vertical dimensions, if it produced a pleasing effect. Landscape architects have the means to move vast quantities of earth, if needs be, and are employed to improve the appearance of actual landscapes, not just their representations; although presenting proposals to clients, committees, and planning inspectors is also an art (perhaps occasionally a shady one).

In Britain, where the Institute of Landscape Architects (now called the Landscape Institute) was founded in 1929, in the heyday of Modernism, practitioners took the creed of functionalism and translated it into a concern for combining use with beauty. The founders of the ILA came from a variety of backgrounds. Geoffrey Jellicoe was an architect who had completed a study of Italian Gardens while studying at the Architectural Association, London. Brenda Colvin (1897–1981), who in 1951 became the organization's first woman president, had trained at Swanley Horticultural College, originally intending to specialize in fruit growing. Thomas Sharp (1901–78) was an up-and-coming town-planner, who would later pioneer ideas of urbanism. The new body's first president was Thomas Mawson (1861–1933), a well-known garden designer who had progressed to town-planning and had already served as president of the Town Planning Institute. The founders dithered for a while over the name, eventually following the American precedent. As many of them had been working on private garden design, there was briefly some thought of having 'Landscape Gardeners' in the title. Colvin later reflected upon what a mistake this would have been: 'it would have taken us much longer to arrive at the full scope the profession has today—if we had arrived at it at all'.

The presence of so many architects and planners in the group ensured that the new institute would not become a coterie of garden designers, but the 'full scope of the profession' only really began to emerge after the end of the Second World War, when the national mood favoured cooperation and reconstruction. The country was wrecked and a returning army expected better living conditions. In this era of post-war consensus, with its measured socialism and Keynesian economics, landscape architects often became involved in large public projects. Significantly, from the 1950s through to the 1980s, the public sector remained the largest employer of landscape architects in Britain. It is only after Margaret Thatcher's neo-liberal revolution that the private sector has employed more, although the Groundwork organization (a charity consisting of numerous local trusts) is now the largest single employer overall.

Among British landscape architects, there is nostalgia for this socially progressive era, because the founders discovered a clarity of purpose which is less often found today, and they were able to influence, not just large scale projects, but also national planning policies. There was also a galvanizing urgency about the problems the country faced, because new housing development, large pieces of infrastructure, and technological developments in farming were rapidly altering the face of the landscape. Many of these issues will seem familiar, not just to British readers who might easily think that we face similar problems today, but also to anyone living in a country which is undergoing modernization, rapid economic development, and landscape change. This makes it worth looking at this period in Britain in more detail.

Agriculture

Agriculture is a good place to begin since many people, when asked to imagine a landscape, will think of fields, farms, and cultivation. Brenda Colvin's *Land & Landscape*, first published in 1940 with a revised second edition in 1970, devoted a chapter to it. The 'humanized, well-lived-in' landscape had an organic beauty, Colvin argued, which could be put at risk by changes of 'policy, use and custom'. The evils of the day were what she called 'suburban spread' (we would now call it 'sprawl' and it will appear again in later chapters), new roads and their associated 'ribbon development', and changes to the system of agriculture. There was, certainly, a class dimension to the panic over ribbon development, and it is interesting to contrast the moral outrage expressed in a book like *England and the Octopus*, architect Clough Williams-Ellis's blast against market forces in development, with geographer and writer John Brinkerhoff Jackson's celebration, in the pages of his own magazine *Landscape* (published 1951–68), of the everyday American landscape of the road, including strip malls, trailer camps and fast-food joints.

In Britain, food security was of concern to the generations who had experienced wartime rationing, but Colvin was worried too about industrial farming, arguing that hedgerows did not need to be ripped out to create efficient farms. 'We too readily discount as "sentimental nonsense"' she wrote, 'any arguments based on the appearance of the landscape, still reacting to the idea of use versus beauty.' In landscape, she asserted, use and beauty are 'fundamentally complementary'. She was not, however, against change, as long as it was thoughtfully considered, which is to say well planned and designed. So, in the middle of a fairly conservative chapter, Colvin suddenly suggests that as far as field patterns are concerned 'we might find that a hexagonal cellular system would be more easily worked, and could provide positions for trees and barns in the angles.' I do not think this idea ever caught on, but it shows a designer thinking through the problem of combining productive efficiency with other virtues that a landscape might possess.

Housing

If unrestrained speculative development was to be brought under control, as indeed it was by post-war planning legislation, then it followed that housing shortages would have to be tackled by the comprehensive redevelopment of poor quality housing stock in cities and the creation of well-planned new towns located in the countryside. The model for this latter form of development had been provided by the town-planner Ebenezer Howard (1850–1928) in his book *Garden Cities of Tomorrow*, who had argued for the creation of a new landscape type which would combine the best aspects of urban living, such as full employment and pleasant society, with the best of country life, such as fresh air and bright homes and gardens, but without the worst features of either: foul air and high rents in the city, or poverty and unemployment in the countryside. The new hybrid was Town-Country and Howard's proposed way of creating it was to build small, self-contained towns, each of no more than 35,000 people, offering jobs and entertainment as well as fields and the beauties of nature. These ideal places were known as Garden Cities and the first two were created at Letchworth (founded 1903) and Welwyn Garden City (founded 1920), both in the orbit of London. These, in turn, inspired the building of a wave of new towns, created under the New Towns Act (1946) and subsequent legislation. There were 11 new towns in the first wave, including Basildon, Essex (designated 1949), Hemel Hempstead, Hertfordshire (1947), Corby, Northamptonshire (1950), and Peterlee, County Durham (1948). A second wave was built between 1961–4, again in response to housing shortages, and a third wave between 1967–70. Five new towns were built in

Scotland, including Cumbernauld (1956) where the landscape architect was Peter Youngman (1911–2005), another member of the pioneering generation.

Landscape architects were involved from the outset. Jellicoe drew up the first plans for Hemel Hempstead based on a variation of the Garden City ideal, in his own words 'not a city in a garden, but a city in a park'. His radical plans were resisted by local people and revised, but he was invited back to design the Water Gardens (1947) where he experimented with the use of symbolism in design. The decorative canal, with its delicate footbridges, was designed in the shape of a serpent, with a fountain where the eye might be and a weir for its mouth. Frederick Gibberd (1908–84), architect, planner, and landscape architect, produced the plans for Harlow New Town, and kept faith in his creation by living there for the rest of his life. Gibberd used the existing topography to structure the town, siting new built areas on the higher ground, separated by open land in the valleys between them. Another pioneering designer, Sylvia Crowe (1901–97), was involved at Hemel Hempstead and Harlow, and then went on to produce landscape plans for Basildon.

The development corporations for later new towns tended to use in-house landscape architects rather than hire outside consultants. Landscape teams under forward-thinking leadership produced innovative ideas. Notably in Warrington, Cheshire, landscape architects set the new development within a structure of woodland and wildflower meadows which brought natural habitats right up to the garden gates of the houses. The designers favoured indigenous species and, in the main, eschewed the planting of exotic ornamental shrubs. This became known in the 1970s as the 'ecological approach' and for a time it was fashionable. Milton Keynes, Buckinghamshire, was a new town large enough to be thought of as a city. Here too there was an ambitiously strategic approach to the landscape plan, with the creation of a linear park system based on river valleys. Different zones of the town were to be differentiated by the character of their planting, so that the town centre would feature horse chestnut, yew, and laurel, while linear parks in the valleys would be replete with willow and dogwood, and the area known as Stantonbury would have lime, birch, and hawthorn. Youngman was involved here too: it was his suggestion that the American-style grid of roads should be softened into a pliable mesh flowing with the landscape. This is not to say that curving streets necessarily make sprawling suburbs acceptable. Developers in the United States have perverted 'townscape' approaches by adopting curvilinear patterns which have no relation to topography at all.

No centrally planned new towns have been built in Britain since the 1970s. Conceived in an era of socialist optimism, faith in technology, and enthusiasm for road building, with the private car seen as liberator, the thinking behind these developments now seems dated and flawed in many ways, although supporters of the new town idea would still say that they are a better way of coping with the seemingly insatiable demand for new housing than leaving it to market forces. The notion of new towns has been superseded by talk of eco-cities, and some countries, including China, have started to build them. Even at the time they were built, Britain's new towns were controversial. When Lewis Silkin, the then Minister of Town Planning, attended a public meeting in Stevenage to announce the designation of England's first new town, protestors greeted him with cries of 'Dictator!' and altered the name of the railway station to 'Silkingrad' to make plain their distaste for central planning. A new town, as much as a new motorway or—to take a topical British example—a high speed railway line, is routinely taken as an imposition by local communities and is usually about as welcome as a meteor strike. One of the tasks which landscape architects have taken on is to minimize the disruption caused by such uninvited developments, attempting to blend and harmonize them with the surrounding landscape. This is also the source of the landscape architect's worst ethical dilemmas:

should a practitioner provide landscape advice to smooth the path of a project of which she disapproves?

Any country undergoing technological modernization is going to need a quantity of new infrastructure: roads, railways, airports, reservoirs, dams, factories, power stations, and more. Fitting all of this into the landscape without damaging valued qualities, such as the historic character of the land, the pleasantness of a view, or the richness of flora and fauna in ancient woodland, is a daunting task, but it is one which landscape architects thought they were particularly qualified to take on. Again focusing on Britain as a case study, there are plenty of examples, across a range of developmental sectors.

Power

The energy supply was, and still is, of pressing concern. In 1963, Prime Minister Harold Wilson gave a speech which is now remembered for his embrace of the 'white heat' of a technological revolution. One of the technologies Wilson admired was nuclear power, which held out the promise of cheap, readily available electricity. But nuclear power stations are inevitably large buildings and because they need to be away from large centres of population and close to a supply of water for cooling purposes, they have tended to be built on coastal sites, in areas which are not densely populated but are often valued for their scenery. Jellicoe was one of the first landscape architects to make proposals for the setting of one. This was the station at Oldbury-on-Severn, mid-way between Gloucester and Bristol, which he described as 'a foreign element that is literally monstrous', in consideration of both its scale and the enormous forces it must contain. This was to be set down in a rural area with an organic pattern of small fields delineated by hedgerows. Jellicoe accepted that there was not much that a landscape architect could do to humanize such a building, but at least he could connect it to its surroundings by designing a landscape which combined the scale of the surrounding fields with the geometries of the reactor buildings. The maquette for the scheme shows a sequence of rectilinear plateaus, reminiscent of the abstract relief paintings of Ben Nicholson, an artist with whom Jellicoe was friendly. Unfortunately, as a result of an underestimate of the soil that would be available, the design could not be completed as intended. Youngman meanwhile became involved with the Central Electricity Generating Board's controversial plans for a nuclear power station at Sizewell on the Essex coast in 1958. Crowe, similarly, was the landscape consultant for Trawsfynydd nuclear power station, built in Snowdonia National Park in 1959–65. Here she designed the area around the monolithic building to blend harmoniously with the surrounding scenery. She wrote a book entitled *The Landscape of Power* in 1958 which dealt not only with the matter of siting power stations but also with ways to minimize the visual impact of the power distribution network on the landscape. She never for a moment doubted that this technological infrastructure had to be accommodated. The dust-jacket of her book stated: 'she accepts the essential need for the construction of immense oil refineries, nuclear reactors, power stations and the network of the electricity grid'.

Dams

Landscape architects were also called upon to mitigate the impact of new dams and reservoirs. Crowe worked on the design for Rutland Water, which by surface area is the largest artificial lake in England. It opened in 1976, supplying drinking water to the densely populated East Midlands. She helped to fit the reservoir into the gently rolling landscape, advised on the siting of ancillary buildings, and made specific proposals to deal with the aesthetic problem of 'draw-down', the exposure of an unnatural looking shore at times of drought. Gibberd was the landscape consultant for another giant reservoir, Kielder Water in Northumberland, which opened in 1982 and is larger in volume than Rutland Water. It was built in anticipation of an expansion of the steel and petro-chemicals industries in Cleveland which never materialized, but it has ensured that the north of England never suffers from a water shortage, and it is now also valued as a scenic and recreational asset. There was much local opposition to the flooding of the North Tyne valley and this may have given Gibberd more sway over the civil engineers in such matters as the shape of the dam, the materials used for auxiliary structures, and even the grass seed specified for sowing on the completed earthworks.

Forests

Kielder Water is surrounded by the country's largest planted forest. This was created by the Forestry Commission which was charged in the 1920s with the task of creating a strategic timber reserve. The single-mindedness of this goal permitted no consideration of aesthetics or ecology, and the commissioners had no brief to develop the recreational potential of the land in their charge. They planted alien conifers, often Sitka spruce, in serried lines, which marched right up to the limits of their ownership, frequently a straight line on the map. When the commission tried this sort of thing in the Lake District in the 1930s there was a storm of protest, but it was not until 1963 that the Forestry Commission began to employ landscape architects to help plan their plantations. The first consultant they engaged was Crowe, who showed how blocks of planting and felling coups could be designed to harmonize with the topography, using natural features to suggest boundaries, rather than basing them on ownership boundaries. The Countryside Act of 1968 required the Forestry Commission to 'have regard to the desirability of conserving the natural beauty and amenity of the countryside'. Henceforth forest managers had to ensure that their forests were not just stockpiles of growing timber, but also that they were attractive and welcoming places for visitors, and were much more varied in their species composition, so that they could support a diversity of wildlife.

Roads

Any major piece of proposed infrastructure is likely to run into opposition. There is often a struggle between central government, which sees an overriding need for the development, and local groups and communities, which seek to defend the values of an existing landscape. Landscape architects often find themselves in the middle of such battles, usually trying to demonstrate that through their mitigation proposals the project need not be detrimental to the prevailing character of the place. In most cases the landscape architect remains a technocentric outsider, although in recent decades there has been a far greater emphasis upon engaging with extant communities. Of all the infrastructure categories landscape architects work on, perhaps road-building excites the strongest emotions. Colvin was one of the first in Britain to work in this area, serving on the Advisory Committee on the Landscape Treatment of Trunk Roads from 1955. In *Land and Landscape* she promoted the American idea of the 'fitted highway'—a road that was harmonized with existing contours, rather than blasted through cuttings or elevated on starkly engineered embankments. She was influenced by the American idea of scenic parkways, which were areas within the 'viewsheds' of highways—the land that could be seen from the road. Landscape in these areas was protected on aesthetic and environmental grounds, a remarkably sophisticated concept considering that these were created in the early decades of the 20th century. Colvin also thought that a well-designed road could enhance a landscape, and she considered the experience of the driver, who needed to be provided with just the right amount of stimulation to be kept alert. She advised against planting trees too close to the highway to avoid the unpleasant flickering effect that could be caused by sunlight falling through their branches. The Highways Agency, which currently manages England's strategic road network, still draws upon advice from landscape specialists in the assessment of new routes and the improvement of existing ones. The aim is still to fit roads to their surroundings and use landform and planting to reduce any adverse impacts upon local landscape character.

Aesthetics and ethics

Of course, if you are hostile to road building on aesthetic or environmental grounds, no well-prepared scheme by a landscape architect is likely to change your mind. Similarly no configuration of artfully contrived earthworks around a reactor building is likely to make anti-nuclear protestors tear up their placards. Individual landscape architects may have crises of conscience when invited to work on proposals for a military airfield or a motorway. One academic opposed to road building labelled landscape architecture 'the night-soil profession' because it was so involved in clearing up the messes left by others. Landscape architects often point out to their critics that controversial proposals are likely to go ahead anyway and that it is better that they should do so with the benefit of a designer. The weakness of such arguments is easily shown up by taking an extreme case, let's say a concentration camp—no amount of aesthetic or ecological finessing could ever make something so morally abhorrent acceptable. This, incidentally, is not such a far-fetched example. As mentioned earlier, many German landscape architects supported the Third Reich. One of them, Wilhelm Hübotter, designed a Teutonic memorial, the Grove of the Saxons, for Heinrich Himmler, which then became a cult site for the SS. After the War he managed to ingratiate himself so well with the victors that he was commissioned to design a memorial for Himmler's victims on the site of the Bergen-Belsen concentration camp in Lower Saxony. Some of the landscape architects who helped to harmonize Hitler's new Autobahnen had also prepared drastic plans for the Germanization of the Polish landscape. It is salutary to remember, once in a while, that landscape architects have not always been on the side of the angels.

Another question that might be asked is whether there is anything dishonest about the screening and camouflaging of infrastructure.

This matter was raised by the Californian landscape architect Robert Thayer in his book *Gray World, Green Heart: Technology, Nature and the Sustainable Landscape.* Thayer argued that we only want to hide our present technology because we are ashamed of it. Hence we try to bury pipes and transmission lines, screen opencast mines, and disguise factories. If we had environmentally sustainable technologies of which we could be proud, he argues, we would want to show them off. Our current technophobia would give way to a happy technophilia. It is an interesting argument and clearly one which involves a cultural shift much broader than anything that can be achieved through landscape architecture alone. The evidence of such a shift is patchy so far. On one hand, there is much cosy enthusiasm for composting and constructed wetlands, on the other, there are bitterly divisive disputes over the siting of wind farms. It seems we have some way yet to go.

Chapter 5
An environmental discipline

Romantics and Transcendentalists

The roots of contemporary environmentalism can be traced back to the Romantics, who turned their backs on an industrializing world to find solace and meaning in nature. A case can be made for claiming that the poet William Wordsworth (1770–1850), who designed his own gardens and those of friends, and whose *Guide to the Lakes* was really a disguised plea for the conservation of a cultural landscape, was also a landscape architect—although the job title had not been invented in his day. Wordsworth was one of the first to identify the problems that picturesque tourism brought in its train. He had celebrated his native Lake District for its beauty and its seclusion, but once it became popular there was little to stop people of means, such as industrialists and merchants from burgeoning industrial cities like Manchester and Liverpool, from building large houses there, destroying the very qualities they had found attractive in the first place. The artist and critic John Ruskin (1819–1900), who also owned a house in the Lake District, railed against a proposal to build a railway line through the area, fearing that there would soon be 'taverns and skittle grounds around Grasmere'. In 1884, Ruskin gave a powerful lecture at the London Institution in which he claimed, through observations made from his home at Coniston, to have detected a new weather phenomenon, the 'plague wind' or the 'storm cloud' which emanated from Manchester, then the most industrialized city in the world. Ruskin later descended into madness, but it is difficult not to see his odd mixture of meteorology and apocalyptic prognostication as a forecast of our present ills: air pollution, global warming, climate change, and extreme weather.

The Romantics were an influence upon the American Transcendentalists, including Ralph Waldo Emerson and Henry David Thoreau, whose writings in turn nurtured the early American environmental movement. The Transcendentalists sought to pastoralize an increasingly technological and urban society. They also believed that the wonders of nature, including spectacular landscapes, were divine and should be treated with respect and awe. Frederick Law Olmsted read and was influenced by Emerson and Thoreau—so much so that one commentator, Lance Newman, has described him as a 'Transcendentalist engineer'. Olmsted was the one who took Transcendentalist ideas and put them into practice. In addition to building pastoral parks in cities, he worked with the naturalist John Muir (1838–1914) to secure the protection of the Yosemite Valley and the Mariposa Grove of redwood trees in California. Through Olmsted an element of environmentalism was woven into landscape architecture from its inception.

Environmentalism

Perhaps it would be better to describe Olmsted as a proto-environmentalist or a conservationist, if only to reserve the label 'environmentalism' for the broad philosophical, social, and political movement which emerged in the 1960s. Another proto-environmental prophet was Jens Jensen (1860–1951), a Danish-born landscape architect who settled in Chicago where he worked for the city parks department before becoming an independent consultant. Observing the spread of urban Chicago, Jensen felt that there was a danger that the inherent character of the Midwestern landscape was in danger of being lost. His contribution to environmentally inflected design was the naturalistic 'Prairie Style' garden which made use of indigenous plants and materials, and drew its form from close observation of the regional landscape. He often included wetland features which he called 'prairie rivers' and 'council rings' which were places for people to gather within the landscape. In 1935, when he was 75, Jensen founded the Clearing in Ellison Bay, Wisconsin, a school which taught a holistic curriculum embracing art, ecology, horticulture, and philosophy.

In 1949, another mid-westerner, Aldo Leopold, published his *A Sand County Almanac.* Leopold, a forester and an expert on wildlife management, was Professor of Game Management at the University of Wisconsin-Madison. He put forward the idea that human beings have a duty towards the land: his much celebrated and often debated 'Land Ethic'. In his formulation: 'A thing is right when it tends to preserve the integrity, stability, and beauty of the biotic community. It is wrong when it tends otherwise.' However it was the publication of biologist Rachel Carson's *Silent Spring* which really raised public awareness of environmental issues. The book showed that chemicals introduced to control insect pests in crops could kill the birds that fed on the insects—the spring was silent because the birds were dying.

The word 'ecology', once attached to a specialized and statistical branch of biology, was soon on its way to becoming the banner for a whole worldview, one which recognizes the complexity and interdependence of the natural world and which, to borrow a phrase from Leopold, 'changes the role of Homo sapiens from conqueror of the land-community to plain member and citizen of it'.

The lessons of connectivity were driven home by the 'Earthrise' photographs taken by the crew of the Apollo 8 mission in 1968. The earth shone like a bright blue marble against the void of space. It looked beautiful, but also vulnerable, and like a voyaging spaceship it had to carry all its life-support systems on board.

A year later, a Scottish-born landscape architect called Ian McHarg (1920–2001), teaching at the University of Pennsylvania, published the most influential book ever written by anyone from the discipline. The book was called *Design with Nature* and it sought to place landscape architecture on a scientific footing. It identified foolishness like building houses on floodplains or among shifting sand dunes, arguing that instead of going against natural processes, we should design with them. As we will see, many of our contemporary ideas about environmentally sustainable design can be traced back to *Design with Nature*. However, McHarg's ideas can themselves be traced back to approaches to design which have emphasized an empathetic approach towards nature, such as the English Landscape School, rather than approaches based on attempts to dominate and control. McHarg was unusual in that his theory had cosmological and metaphysical dimensions, but could be distilled into a step-by-step method for landscape planning, which began with a detailed survey of such things as geology, soils, climate, and hydrology.

Environmentalism was born in protest against the harm done through human industry and exploitation of industry. Back in 1991, the environmental scientist Tim O'Riordan drew a distinction between 'technocentrist' and 'ecocentrist' environmentalists. The former were optimists who felt that existing economic and social arrangements were capable of dealing with environmental problems, whereas the latter, who included Deep Ecologists, Gaianists, Communalists, and Red-Greens, held that some redistribution and decentralization of power was necessary. This more radical strand has emerged again in recent anti-Capitalist demonstrations. Whatever the personal views of individual landscape architects, it is clear that the practice of landscape architecture, taken as a whole, is more of the managerial kind. The underlying belief is that human relations with nature can be improved through planning, design, and management, not that the world first needs a revolution.

Although McHarg is often referred to as a rigorously ecological thinker, ultimately he too was advocating reform rather than revolution and an eventual accommodation between human beings and nature. *Design with Nature* was a textbook for this kind of work.

Bioregionalism and the locally distinctive

Many landscape architects have embraced the notion of bioregionalism, which is akin to environmentalism. The term, which was coined by the counter-cultural activist Peter Berg in the 1970s and promoted in the 1980s by the journalist Kirkpatrick Sale, refers to a movement which shares environmentalism's aspiration to live in harmony with nature, but which puts great stress upon the local. Bioregions are defined through their physical and environmental features, including their soils, flora and fauna, landscape characteristics, and watersheds. Although cultural factors are also important, bioregions are not defined by political or administrative boundaries. For some environmentalists, humanity can seem to be the enemy, but bioregionalists see humans as residents of bioregions, and they work to reinforce the connections between human societies and place. Needless to say, this ideology stands in blunt opposition to those globalizing trends in advanced capitalism that tend to make everywhere more like everywhere else. In *LifePlace* the Californian landscape architect, Robert Thayer, reflected on what bioregionalism might mean for everyday living and explored the possible social benefits of re-inhabiting the natural world on a local scale. His book was part memoir, part lifestyle guide. It suggested that we should learn to connect with our local surroundings, living closer to the land, eating locally produced food, and living in houses designed to fit their regional context.

In general, landscape architecture has always elevated and celebrated that which is locally distinctive. This stems, I believe, from the requirement to 'consult the *genius loci*' which derives from a Classical tradition that special places had their own local deities, such as *naiads* and *driads*, but really means 'pay attention to the existing qualities of the site'. Using local materials and indigenous plants is, as Jensen saw, a pathway towards harmonious design which respects characteristics of place. In the Netherlands, Thijsse strongly criticized existing park practices for distancing humans from nature. He argued that a new type of park was necessary, one which could make people aware of the richness and diversity of their local landscapes. This could be achieved, he argued, by bringing the flora and fauna of the countryside into the town for everyone's edification and enjoyment. He promoted the idea of the 'instructive garden' building the pioneering example at his own home in Bloemendaal. Later J. Landwehr, the Director of Parks for Amstelveen, a dormitory suburb of Amsterdam, used the term *heempark* (home park) to denote parks which predominantly featured native wild plants. Landwehr created what is still the best known example by creating a waterside park and naming it the Jacques P. Thijsse Park in honour of the innovative botanist. The home parks were a significant influence upon landscape architects around the world. Dutch ideas about native planting were introduced to Britain by Alan Ruff who taught at the University of Manchester. His ideas were widely taken up in the 1970s and 1980s, particularly, as we saw in Chapter 4, in some of the English new towns, and it became customary to refer to the 'ecological approach'. In many ways this chimed with Modernism, because plants were to be selected not for their showy flowers or shiny leaves, but for the role they would play in an ecosystem. The argument was that native plants were plentiful, cheap, and easy to establish and maintain. Native trees, such as alder, willow, birch, ash, and oak, could be planted in large numbers to create 'structural woodlands'. Formal qualities were not of much concern, indeed the approach was almost anti-design and there was a strong belief that it was the users of these landscapes who would ultimately determine their form. As such plantations grew, the benefits they provided, which included shelter from winds, opportunities for recreation, support for wildlife diversity, and resources for education, would all increase, while the cost of managing the woodland would get less, something that could not be said for ornamental plantings in manicured parkland. Many of the techniques developed during this period, which included methods for establishing species-rich meadows and wetlands, soon became landscape architectural stock-in-trade, so much so that the idea of a distinctly 'ecological approach' lost much of its meaning. However, it also prefigured many contemporary ideas, such as green infrastructure, ecosystem services, and even landscape urbanism, which I will discuss in greater depth later on.

Landscape ecology and ecosystem services

The next major development was the emergence of landscape ecology in the late 1980s. This was an understanding of ecology at the scale of the landscape. It emphasized pattern and process, and many of its key concepts, such as matrix, patch, corridor, and mosaic, are spatial. A 'patch', for example, might refer to a wood, a meadow, or a marsh; a 'corridor' could be the banks of a river or even the verges of a motorway. In *Land Mosaics: The Ecology of Landscapes and Regions*, Harvard ecologist Richard Forman hypothesized that 'for any landscape, or major portion of a landscape, there exists an optimal spatial arrangement of ecosystems and land uses to maximise ecological integrity'. In short, planning and design were not just matters of aesthetics or amenity. The way that landscapes were laid out could make a difference to the way they functioned ecologically. To take a simple example, a major new highway cut through a woodland could isolate a fragment of the wood and, in the event of numbers of a particular species declining, it would be hard for the species to repopulate the disconnected patch. Landscape ecology provides the scientific understanding to back up our intuition that the fragmentation of habitats through the expansion of human development is a bad thing. Thus, when landscape architects are asked nowadays to give advice on a development project, their brief will not be limited to aesthetics or the convenience of the proposals for human beings, but they will also have to consider its consequences for habitats and ecosystems. Through their design they may seek to maintain or improve the existing levels of ecosystem connectivity. Fortunately many of the features which favour species diversity are also those which appeal to humans, such as large parks, wooded riverbanks, or footpaths constructed along disused railway lines. In addition, landscape architects have become proficient at translocating habitats. A mature hedgerow, for example, can be carefully dug out and replanted in a different location. Species-rich grasslands can be lifted and transferred to carefully prepared receptor sites. Such methods are sophisticated echoes of the techniques used by 18th-century landscape improvers like 'Capability' Brown, who would often translocate mature trees to create more pleasing views for their wealthy land-owning clients. They are sometimes criticized, however, on the grounds that transposed habitats do not thrive as they might have done if left in place.

Although many environmental philosophers have attempted to defend the natural world by asserting that it has an intrinsic right to exist, it seems that arguments based on human wants and needs are usually more persuasive. These arguments are labelled 'anthropocentric'. They include the argument that nature is the source of many aesthetic and spiritual satisfactions. Rather more pressing perhaps is the thought that without the intricate web of nature and the multitudinous contributions of a sweeping range of living things, human life itself would not be sustainable. One formulation of this argument is found in the notion of ecosystem services. In many respects, this has been understood since at least the time of Plato, who warned about the perils of deforestation and soil erosion in his book *Critias*, but it appeared in its present expression with the publication in 2005 of the United Nations Millennium Ecosystem Assessment, a four-year study involving more than 1,300 scientists worldwide. The problem has been that many of the services which ecosystems provide have appeared to be free. Placing a monetary value on them allows them to be entered into economic calculations. There can even be a market in them, although some environmentalists baulk at this neo-liberal way of thinking. New York City, for instance, pays for water services in the Catskills and Delaware catchments, and this is considered a good deal when compared with the cost of building and running water purification plants. The services are extremely extensive, ranging from the pollination of crops by bees, to the provision of foodstuffs and medicines, carbon sequestration, the purification of water and air, and the decomposition of wastes, but also non-material benefits such as places for relaxation, recreation, and spiritual uplift.

The concept of ecosystem services is potentially very significant for landscape architects and landscape planners. For much of its history, landscape architecture has struggled to throw off the notion—associated, no doubt, with its origins in landscape gardening for an elite clientele—that it is a discipline mostly concerned with taste and aesthetics, and thus something superfluous or superficial. Unsurprisingly, landscape architects have never thought this, indeed it is for many a passionate vocation, but the message about the importance and centrality of the discipline has sometimes proved difficult to convey. However, if it becomes well established that ecosystems provide services, and that these services can valued at astonishing sums, and if it also becomes evident that these ecosystems are embedded in the landscapes we inhabit, then the services of landscape architects should be more in demand than they ever have been.

Regenerative design

The idea that landscapes can do things for us was captured in the idea of 'regenerative design' advanced by John Tillman Lyle (1934–98) who was a professor of landscape architecture at the California State Polytechnic University, Pomona. Lyle was the author of *Regenerative Design for Sustainable Development* and the principal architect for the Center for Regenerative Studies at Pomona, where a community of faculty and graduate students developed a community which produced its own food and energy and treated its own wastes, thus demonstrating that it was possible to live within the limits of available resources without causing environmental degradation. Lyle drew a distinction between two ways of living, which he called 'degenerative' and 'regenerative'. Degenerative living uses up limited resources and fills natural 'sinks', such as the atmosphere, lakes, rivers, and the oceans, with damaging waste products. It is a linear process, a 'one-way throughput system' heading toward a dystopian future. Regenerative living, on the other hand, provides for the continuous replacement of the energy and materials through forms of recycling. Lyle shows how the landscape can be modified to incorporate regenerative systems. Infiltration basins can be constructed above aquifers to help them to recharge. Solar collectors can be positioned where incident radiation is high. Grey-water from activities like doing the laundry, dishwashing, and bathing can be reused for irrigating crops. Lyle's books are bursting with suggestions, and many of these 'neotechnologies' (to borrow his phrase) have been put into practice at Pomona.

Many of the technologies collected by Lyle have now found their way into mainstream landscape architecture practice. A good example would be the design of sustainable drainage systems (SuDS), sometimes referred to as Water Sensitive Urban Design. In traditional drainage systems, water is channelled away from the site in pipes and sewers. In many old urban systems sewage and storm-water share the same conduits and this can have unpleasant consequences if capacity is exceeded. Manhole covers blow off and the streets are treated to the delights of faecal fountains. As global warming disrupts weather patterns, deluges of rain and the associated damage from extensive flooding have become more common. Sustainable drainage systems use vegetated swales (broad ditches) and filter strips to slow down runoff, while permeable surfaces and infiltration devices such as soakaways, rubble drains, and infiltration basins help water to

6. This golf course in Durango, Colorado, won an Honour Award from the American Society of Landscape Architects in 2007. It incorporates a hierarchy of constructed wetlands and swales to collect and purify water before it reaches existing wetlands and streams

percolate into the ground, reducing the risk of flooding (Figure 6). The principle is to dispose of water, as far as possible, on site, rather than to pipe it somewhere else. This is characteristic of many regenerative technologies, they are small scale but widely distributed. If the aesthetic problem which challenged 20th-century landscape architects was how to accommodate a relatively small number of giant dams and massive power stations, the challenge now is how to site hundreds of thousands of wind turbines and solar panels.

A concept which brings together many of the ideas explored in this chapter, particularly those of landscape ecology, regenerative design, and ecosystem services, is that of 'green infrastructure planning'. Discussion of this must be deferred to a later chapter on landscape planning more generally, but the essential notion is that a network of green spaces, whether semi-natural or designed, delivers benefits analogous to those delivered by the road network, the sewerage system or the electricity grid. Public parks, green roofs, village greens, canal banks, community gardens, and allotments (to name just a few examples from a wide typology) can all be considered components of green infrastructure. And since these are just the sorts of place which often concern landscape architects, it is perhaps not surprising that green infrastructure is one of the discipline's current enthusiasms.

Chapter 6
The place of art

An art or a science?

For several years I was in the position to select students for a Master's programme in landscape architecture which led to a professional qualification. This taught me that it is very difficult to predict who will make a good landscape architect or what sort of previous accomplishments might indicate potential. I looked out for an interest in places and some indication that a student might be able to think spatially. It was certainly helpful if they could produce evidence of an ability to draw. Beyond this, the subject or class of a first degree was little guide to performance. We recruited a lot of geographers, a number of architects, a quantity of botanists, ecologists, environmental scientists and horticulturists, and a smattering of fine artists. One interesting category consisted of students with a background in science who had enjoyed art at school but had been forced to abandon it through the constraints of the syllabus. They often proved to be good students, for whom landscape architecture offered the perfect outlet for their abilities. Usually it was a discipline they had not heard of while at high school.

The British school system and perhaps the structure of education in most parts of the world usually forces students to make an unwelcome choice between the arts and the sciences, a decision which often shapes their whole lives. It is rare to find students studying a 'hard' science, such as biology, physics, or geology, at the same time as pursuing courses in painting, photography, or graphic design. One of the most appealing aspects of landscape architecture is that it sees this kind of transdisciplinarity as a virtue. Some of its practitioners are genuine polymaths and most are at least generalists who can understand a report from an ecologist as well as they can see meaning in a painting by Constable or Cezanne. Social and political awareness is important too, and should be addressed in the landscape architecture curriculum. But if landscape architecture is a broad church, this does not mean that individual landscape architects are without their own inclinations and prejudices. The corny old question 'Is landscape architecture an art or a science?' has been kicked around in many a seminar, but it still divides opinion. There are those who, like McHarg, prefer to see landscape architecture as applied ecology and those, conversely, who see it primarily as a form of art, regard designed landscapes as carriers of meaning, and see the landscape as a medium for expression.

Of course, the word 'art' is difficult to define. At its broadest, it can mean something like 'skill' or 'craft'. Olmsted often used the word this way: after visiting Birkenhead Park he remarked that 'art had been employed to obtain from nature so much beauty'. However, Olmsted also suggested that the science of the engineer was 'never more worthily employed than when it is made to administer to man's want of beauty. When it is carried into works not merely of art but of fine art.' This is an interesting assertion, not just because Olmsted was placing engineering and science at the service of art, but also because he was clearly saying that art is more than just a matter of skill and that it has some connection with beauty. Olmsted was much influenced by English aesthetic ideas from the 18th century, when landscape gardening had been the sister art of painting and poetry. Although we often describe landscapes as beautiful, it would be anachronistic to follow Olmsted in associating fine art with beauty. As the art critic Arthur Danto has suggested, the Modernist avant-garde dethroned the pursuit of beauty as the principal goal of art, replacing it with the purpose of embodying meaning—which is not to say that art cannot be beautiful or that beauty does not matter in our everyday lives (it clearly does), nor that landscape architects should not be concerned with it, but when thinking about the artistic possibilities available to landscape architecture, it is not necessarily the best starting point.

Jellicoe's theory

Geoffrey Jellicoe, a staunch believer in landscape architecture's credentials and mission as a form of art practice, saw this too. As we saw in Chapter 3, Jellicoe believed that landscape architecture is at its strongest when it has a firm connection with the fine arts, particularly (for him) the art of painting. For Jellicoe, landscape architecture's mission was not merely to arrange things tidily or to clear up visual mess—that was just the pursuit of seemliness. The landscape architect's higher calling was to create landscapes that were 'as meaningful as painting'. Jellicoe had his own theory about the way landscapes could mean, although it is hard to find anyone who believes it today. Influenced by the analytical psychologist C. G. Jung, Jellicoe argued that by spending time on site a designer could tap into the 'collective unconscious', a sort of psychic substrata shared by all human beings. The subsequent design would embody universal archetypes, and these could have a powerful, but largely unconscious effect upon visitors to the landscape. The theory is mystical and untestable, but it sits comfortably with an experience that most of us have from time to time, which is the sense that a place has a powerful presence or atmosphere.

Most commentators agree that landscapes can be meaningful, but the extent to which meanings can be 'designed in' is often debated. The landscape architect Laurie Olin, who received the 2012 National Medal for Arts from President Obama and whose celebrated designs include the revamping of Bryant Park (1992) and Columbus Circle (2005) both in New York City, wrote an article in 1988 entitled 'Form, Meaning and Expression in Landscape Architecture' which was provoked by his sense that his discipline had fallen under the sway of the 'born-again language of fundamentalist ecology'. Meaning came back into fashion, and tutors would earnestly demand that their students should explore metaphors and explain their concepts. This in turn prompted Marc Treib, Professor of Landscape Architecture at the University of California, Berkeley, to wonder 'Must Landscapes Mean?' (the title of an essay he wrote in 1995). Treib suggested that attempts to build in meaning from the outset often backfired and that designers should concentrate on making places which give pleasure. If designed places become popular, then meanings will accrue.

Isamu Noguchi and the borderlands between art and design

Although art and design are often bracketed together, there is a divide and most practitioners know which side they fall. A landscape architect once told me that he did not aspire to create art—his goal was to produce 'good design'. Similarly a sculptor who placed work in public places said that he did not produce work to order—that would be 'mere design'. However, there have been practitioners who defy easy categorization, such as the Japanese American Isamu Noguchi (1904–88) who trained first as a sculptor under Constantin Brancusi, but began submitting proposals for public spaces and civic monuments in the 1930s. He is as often described as a landscape architect as he is described as an artist, on the strength of some high profile garden commissions including the grounds of the Connecticut General Life Insurance headquarters in Bloomsfield, Connecticut (1956), a Peace Garden for the UNESCO Building in Paris (1956–8), and the Cullen Sculpture Garden at the Museum of Fine Arts in Houston, Texas (1984–6). Noguchi was a modernist whose work was influenced by

Japanese traditions, most notably in the Marble Garden for the Beinecke Rare Book Library at Yale University, New Haven, Connecticut (1960–4), which features a low pyramid, a cube balanced on a point and a standing ring, all in white marble and without a plant to be seen, referencing the Zen tradition of dry gardens (*kare-sansui*) exemplified by the celebrated temple garden at Ryoan-ji. Noguchi also designed gently contoured children's playgrounds with equipment that would have looked equally at home in a sculpture gallery. In effect, these proposed playgrounds would themselves have been large-scale bass-relief ground sculptures. Noguchi was able to collapse the distinction between 'functional' design and self-directed artistic practice, and his work has been influential in both spheres, even though he had great difficulty in persuading the authorities to build any of his playscapes. Only two Noguchi playgrounds were constructed in his lifetime. One is the Kodomo No Kuni (Children's Land) Playground near Tokyo, which he built in collaboration with Yoshio Otani in 1966. The other, opened in 1976 and restored in 2009, is in the Olmsted-designed Piedmont Park in Atlanta, Georgia.

It is sometimes said that artists pursue self-imposed goals and seek answers to self-posed questions, whereas designers respond to a brief, work in the context of a need, and have the eventual users of their designs in mind throughout the design process. This is broadly true, but Noguchi's eclectic and synthesizing practice shows that there is nothing hard and fast about these distinctions. The border between art and design can be porous.

Landscape and land art

With his interest in the earth as a medium for sculpture, Noguchi was a decade or two ahead of his times. The art form with the greatest influence upon landscape architecture in the latter half of the 20th century was not painting, as Jellicoe had thought, but sculpture, or at least the particular movement variously known as Land Art, Earth Art, or Earthworks. This emerged in the 1960s and 1970s, with its origins in Conceptual Art and Minimalism. It was also a specific response to the commercialization of art in the gallery system of the time. By choosing to make their works in remote places like the deserts of Nevada, New Mexico, or Arizona, land artists such as Robert Smithson (1938–73), Michael Heizer (1944–), and James Turrell (1943–) turned their backs on the galleries, though not necessarily upon the wealthy patrons and foundations that supported their work. Land Art shares certain characteristics with landscape architecture. It is generally 'site-specific', which is to say it can only be made in the place it is sited. Like works of landscape architecture, works of Land Art are responsive to the character or *genius loci* of the place in which they are created. Usually they are also made of the stuff of that place. As with landscape architecture, they can involve large scale earth moving. For instance, Smithson's *Spiral Jetty* (1970), one of the best-known examples of the genre, was built on the shore of the Great Salt Lake, Utah, using basalt rocks and mud. It is 460 metres long and 4.6 metres wide and it is now encrusted with salt crystals. Heizer's *Double Negative* (1969) is a trench, 9 metres wide and 15 metres deep, which straddles a natural canyon in Nevada. Some Land Artists became involved with land reclamation after mining, another incursion into the physical and conceptual territory of landscape architecture. Heizer was commissioned by the Ottowa Silica Company Foundation to create a series of *Effigy Tumuli* at Buffalo Rock, Illinois. These works (completed 1985) draw upon Native American traditions of mound building and each represents a creature indigenous to the region: a catfish, a water strider, a frog, a turtle, and a snake. An up to date example of similar work is the *Northumberlandia* landform sculpture (2012) created by the architect/artist/critic Charles Jencks for a former open-cast coal mine near Cramlington, Northumberland. Artists who have associated themselves with open-cast mining have often become mired in environmental controversy, as they can be seen as aiding and abetting a destructive industrial operation, a charge that is also sometimes levelled against landscape architects who work in this sector.

Land Art implied no particular brief for nature, but some of its early practitioners were nevertheless interested in ecology and environment. Alan Sonfist's *Time Landscape* (1965–present) consisted of a rectangular plot in Lower Manhattan which the artist planted with species that would have grown there in pre-colonial times. The area is managed by the city's parks department as a developing woodland and is regarded as a living memorial to the forest that once covered the island. Newton and Helen Mayer Harrison (often referred to as 'the Harrisons') are pioneering ecological artists who have become involved in causes such as watershed restoration, urban renewal, and responses to climate change, which might usually be seen as the domain of environmental professionals such as planners or landscape architects. For instance, a recent installation, *Greenhouse Britain 2007–2009* suggests ways in which people might withdraw from low lying land as sea levels rise. Land Art was succeeded by Ecological Art or Environmental Art, whose practitioners were motivated by ethical concerns about the consequences of human activity upon the planet. Some of this work shades into landscape architectural practice and there have been successful collaborations between artists and landscape architects. One of the most notable was that between artist Jody Pinto and landscape architect Steve Martino at *Papago Park* (1992), on the border between the cities of Scottsdale and Phoenix, Arizona. Martino has pioneered the use of indigenous, drought-tolerant plants for landscape work in the American south-west. He and Pinto created a water-harvesting structure which, when seen from above, resembled the branches of a tree. By detaining rainfall and allowing it to percolate, this design aided the regeneration of the flora on the site, including the characteristic saguaro cactus.

Landscape architecture's avant-garde?

Regardless of the influence of architectural Modernism upon Halprin, Eckbo, Kiley, and their ilk, the philosopher Stephanie Ross has suggested that landscape architects and garden designers missed the avant-garde opportunities that were seized by other disciplines. Where are the garden equivalents of John Cage's *4'33"*, the musical composition which consists of four minutes, thirty-three seconds of silence, or William Burroughs' literary cut-ups, writings that could be reassembled in any order? This sort of introspective attention to the materials and processes of the medium characterized the avant-garde. Ross tried to imagine what an avant-garde garden might look like—perhaps it would consist of a display of garden hoses?—before concluding that garden designers had baulked at the challenge and that Land Artists and their successors had stepped into the cultural vacuum this had created. Of course, for those who conceive of landscape architecture as design or as planning, the lack of an avant-garde is unproblematic.

In any case, Ross's account, which is to be found in her book *What Gardens Mean*, oversimplifies the relation between Land Art and landscape architecture/garden design. There are examples of gardens which push the boundaries of what can be thought of as a garden, and some of these have been created by landscape architects. Martha Schwartz horrified those with settled opinions by using unconventional materials and fake plants in her projects. Coming from art practice, she launched her new career with the witty Bagel Garden (Boston, MA, 1979) which used shellacked (varnished) bagels as decorations in a domestic parterre (a parterre is a design on the ground, usually edged with box hedging and filled with coloured earths or gravel) which poked fun at its grander equivalents in French formal gardens. It was a pivotal work which caused a rift in the discipline at the time. Stella's Garden (Bala-Cynwyd, PA, 1982), which Schwartz created for the yard of her mother's duplex, employed chicken wire, netting and shards of Plexiglass in a design which required no gardening skills whatsoever to maintain. On the rooftop of the Whitehead Institute, Cambridge, MA, Schwartz designed the Splice Garden (1986) which apparently combined a Renaissance garden with a Japanese garden, but none of the plants were real. The clipped hedges were made of steel covered with astroturf and a faux topiary bush 'grows' horizontally from one of the enclosing green walls. Some landscapes architects harrumphed—this might be art, but was it really landscape architecture?

From these small beginnings, Schwartz built a large international practice, winning commissions for important public spaces in major cities, but she did not lose her provocative edge. Martha Schwartz Partners' plans for the redesign of Exchange Square, Manchester, UK (completed 1999), ran into trouble with local politicians because she included five ersatz palm trees, a dig at the city's reputation for grey skies and rain. They were replaced by windmills in the final scheme. Her design for Grand Canal Square, Dublin, Ireland (completed 2008), is an upscale descendent of Stella's Garden with a red carpet made of resin and glass and an array of tilted red poles which are illuminated at night. Schwartz's work is often visually arresting and she seems to carry the day with design juries, but she is sometimes criticized for not working in a more consultative way. The Hall of Shame website maintained by the Project for Public Spaces has featured some of her high-profile projects, including Exchange Square, Manchester, UK, and Jacob Javits Plaza (sometimes referred to as Federal Plaza/Foley Square) in downtown New York City. Schwartz's design for the latter featured looping green benches around hemispherical green mounds, and it won an Honour Award from the American Society of Landscape Architects in 1997. Critics, however, said that the pop-art treatment of the space failed to provide comfortable places for the workers from the surrounding office blocks to use, and it is perhaps significant that, at the time of writing, the square is being reconfigured by Michael Van Valkenburg Associates. But perhaps it is harsh to single Schwartz out for criticism. The Project for Public Spaces often criticizes renowned landscape architects for what it sees as self-indulgent design which creates dramatic imagery but not vibrant civic spaces. Schwartz has gone from *enfant terrible* to grande dame of the discipline and she now shares her role as provocateur with her protégé Claude Cormier, whose Boules Roses were discussed in Chapter 2. Cormier shares Schwartz's irreverent sense of fun. His project for the Four Seasons Hotel in Ontario, Canada (2006–12), features a 12-metre high cast iron grand fountain, looking something like a giant cake-stand, along with the pixelated arabesques of an out-of-scale 'urban carpet' made from granite blocks.

The empathetic approach

Land Art at its most assertive can be seen as an imposition upon landscape, but there is a quirkier tradition, exemplified in Britain by artists such as Richard Long (1945–) and Andy Goldsworthy (1956–) and in Germany by Nils-Udo (1937–), which involve restrained and often delicate interventions in place. Much of their work eschews permanence and monumentality. This work is often much admired by landscape architects, because it seems to share the discipline's concern for the *genius loci* or particularity of place. Trudi Entwistle, who teaches landscape architecture at Leeds Metropolitan University, UK, is also a site-specific artist in this mould. She has made the opposite journey to Schwartz, training first as a landscape designer but becoming an artist. She says that her work 'lies somewhere between the boundaries of land art, sculpture and design. It is site specific and investigates how sculptural forms integrate with their surroundings, interacting with human movement and the changing elements of light, weather, natural growth and decay'. Her work does not set out to compete with the landscape but in some way to complement it. They are subtle additions which one might stumble upon (Figure 7). Some, like *Drift* made for the Busan Biennale, Busan, South Korea (2002), and *Wave Break*, Guisseny, France (2007),

7. Trudi Entwistle's 'Apple Heart' (Turku, Finland, 2008) is situated in the grounds of 'Life on a Leaf', a fabulous house inspired by the forms of nature, created by artist Jan-Erik Andersson (built 2005–9). It was inspired by a Finnish love story, the 'King and the Castle', which also inspired the leaf house

are pieces which provide people with some temporary affordance such as shelter from the wind, but functionality is not their primary purpose.

Art as possibility not imperative

If we ask whether landscape architecture can be art, we can easily get ensnared in all kinds of muddle. We often use the word 'art' in a laudatory way, rather than as a descriptive term for a sort of human activity. In the laudatory sense, there can be no such thing as bad or indifferent art. 'Architecture' gets used in this way too (sometimes with a capital 'A') to mark out a special category of structures which transcend mere building. It is also a moot question whether Architecture can be considered Art, or whether the practical purposes it must serve somehow get in the way. In any case 'landscape architecture' is not generally used in this evaluative way, partly, I think, because it is a relatively recent term and partly because the range of activities that must squeeze under its umbrella includes so many things that owe more to rational planning than to creativity, such as environmental impact assessments or zone of visual influence analyses. We do not say: 'this park is Landscape Architecture, but that one is just a designed landscape'. Nevertheless, there is a generally agreed canon of great works, including such masterpieces as Ryoan-Ji, the Villa Lante, the parks of 'Capability' Brown, Central Park, and Thomas Church's garden for the Donnell Residence. Do works of this stature deserve to be called art? I think the answer has to be yes, but it does not follow that the motivating purpose of landscape architecture must be the creation of works of art. If Treib is right, then packing designs with allegories and allusions may not be the royal road to significance, in any case. It may be that making places which stir the emotions and give pleasure is a surer purpose to pursue. Indeed, even the pursuit of mere seemliness, contra Jellicoe, might be satisfying enough for many a practitioner and many a client.

Chapter 7
Serving society

In 2008, Martha Schwartz took part in a Channel 4 TV programme called 'Big Town Plan', presented by architect Kevin McCloud, in which designers were invited to make improvements to spaces in Castleford, a former mining town in the north of England. Schwartz was invited to design a 'village green' for New Fryston, a community on the outskirts of the town. She did not see eye-to-eye with local residents, but her scheme was built anyway. According to an article published later in *Horticulture Week*, the locals nicknamed the sculpture which the American placed in the centre of the green 'Martha's Finger', reflecting their feelings about her way of working. Phil Heaton, another landscape architect involved in the programme, told the magazine, 'Martha Schwartz is a wonderful designer but a bit of a prima donna... Designers need to listen before trying to impose ideas and that is where Martha Schwartz went wrong.' On the other hand 'design by committee' is something which makes most landscape practitioners nervous, associating the phrase with bad decisions and uneasy compromises. The landscape *auteur*'s clarity of vision will be watered down by those who can't or won't see it. Those on the artistic wing of the discipline probably feel this more keenly than others, but there is also a strong counter argument which says that landscape design that ignores users' wishes is bad design.

Landscape, power, and democracy

The architecture critic Rowan Moore has written that 'Architecture is intimate with power. It requires authority, money and ownership. To build is to exert power, over materials, building workers, land, neighbours and future inhabitants.' I fear that this is true, even though we might look for examples of the power of the collective, barn-raisings in rural America, for example. A similar statement could be made about landscape architecture, at least if we are considering the canonical designs that are found in history books. These show that it took surplus wealth to create parks and gardens, and for most of the time this was in royal or privileged hands. This did not necessarily make life easier for landscape designers, but the issues were different. When André Le Nôtre laid out the vast gardens of Versailles for Louis XIV, he had to contend with court rivalries, the changeable ideas of an absolute king and a degree of interference from royal mistresses, but on the whole he knew who his client was and what would please him. He did not have to worry much about anyone else. Designed landscapes, such as the gardens of Versailles, were expressions of mastery and control. English landscape parks of the 18th century certainly looked very different, but they were about the display of wealth and power too. Power came through control of the land, and wealth was needed to employ the labourers and horses required for necessary damming of rivers and recontouring of the ground. These parks were often created for men who talked much about British liberty, but it was their own freedom from arbitrary royal power that concerned them. They were not generally on the side of the common man. Infamously, when Joseph Damer, who later became the Earl of Dorchester, employed 'Capability' Brown to work on his estate at Milton Abbas, he asked him to relocate the villagers who were his neighbours to a new settlement built half a mile from his great house. One stubborn inhabitant would not go, so Damer ordered Brown to flood him out. For most of history, the only listening landscape designers had to do was to their paymasters.

The democratization of landscape architecture began with the 19th-century movement for public parks. Now the client was a public client, generally a council of elected representatives, and the parks' users were the citizenry in all their complex diversity. Part of the brief was to provide a park which would appeal to all classes, and behind this often lay the paternalistic hope that such social mixing would reduce tensions within society. The revolutionary zeal of early Modernism took things further, placing a social mission at the heart of the design enterprise. The German Bauhaus (1919–33) was founded around ideals of socialist design and production. The utopian idea that a rational, functional architecture, pre-fabricated and mass produced, could improve living conditions for all was adopted in many countries, not least in Britain, where Nye Bevan, the 1945 Labour government's first Housing Minister, declared that nothing was to be too good for the working man. Politicians and planners put their faith in high rise blocks of flats but the dream soon soured, and many British tower blocks met the same fate as the Pruitt-Igoe flats mentioned in Chapter 3. There were, however, some significant successes, such as Ralph Erskine's Byker Wall in Newcastle (1969–81), which achieved a sense of place through attention to orientation and topography. Landscape architects were involved in the design and planting of the communal spaces in the low rise development sheltered by the Wall, and notably the design team consulted with the residents of the terraced houses in the old Byker who would become the tenants of the new social housing.

Modernist housing did not have to be high rise; two British architects, Eric Lyons and Geoffrey Townshend, teamed up with landscape architect Ivor Cunningham (1928–2007) to form Span Developments Ltd, a company which built Modern suburban homes in Kent, Surrey, and East Sussex, re-animating the ideals of the garden city movement and incorporating large communal gardens to the front of the properties. Both the Span housing and the Byker estate have a northern European sensibility and it is easy to draw parallels with housing projects in Scandinavia, such as the Søndersgårdparken development at Bagsværd in Denmark (the architects were Hoff and Windinge; the landscape architect, Akserl Andersen, 1943–50), where rows of low rise housing are grouped around a large community green flanked by tall poplar trees. In all of these schemes there was a conscious attempt, on the part of the designers, to foster sociability through the design of open space.

Empathy

If a landscape architect is going to serve people—and it is difficult to think of a project that does not involve this to some degree—then a capacity for empathy is required, and this involves a kind of imagination, the ability to place oneself in the shoes of another, no matter how different that person might be in terms of life experience, and physical and psychological characteristics. Since it takes a prolonged period in higher education to qualify and since the profession is comfortably middle class, there may already be little overlap between the life-worlds of the designers and those of many of the people they design for. It is not the case, however, that the great majority of landscape architects are men. The website of the New Zealand Institute of Landscape Architects, for example, observes that while (building) architecture and landscape architecture both start out with an even balance of genders at university, only 18 per cent of registered architects are women, whereas the corresponding figure for landscape architecture is 42 per cent. A recently published American book, *Women in Landscape Architecture: Essays on History and Practice*, edited by Louise A. Mozingo and Linda L. Jew, observes that the realm of landscape provided women with an alternative to domesticity, and that it has always been more receptive to female practitioners than has been architecture, engineering, or science. If there is a specifically gendered way of experiencing the landscape, then the presence of so many female practitioners ought to ensure that it is represented in design practice and that women's concerns are an influence upon the places that are created. A positive example might be the attention given to matters of safety and fear of crime in public parks, where placing tall, dense planting adjacent to footpaths is usually avoided, lighting levels are carefully considered, and alternative routes are included to provide egress in an emergency.

Empathy is a good thing, but perhaps it has its limits. At one of the landscape architecture schools where I have been an examiner, it was customary to set aside some time each year for able-bodied students to become wheel-chairs users for a few hours, to understand how they coped with various surfaces, changes of level and ramps around the campus. Similarly, sighted students were asked to wear blindfolds and use canes. Recently there has been a flood of designs for sensory gardens, supposedly with the visually impaired in mind, but the sense of sight and the communication of ideas through drawings have been so central to the landscape enterprise that I have not, in 30 years of practice and teaching, encountered a visually impaired student or practitioner. Considering such forms of difference helps one to understand the limitations of empathy. It is easy to give people, not what they want, but what you *think* they want, or need, or ought to want. Most designers make this mistake at some time. I remember designing a very elaborate play-fort out of logs for an open space near a council estate in Gateshead, Tyne and Wear. It looked terrific on the drawings, and indeed when it was built, but its curtain walls provided the ideal concealment for teenagers sniffing glue and the whole thing had to be dismantled. My fault was that I had not asked anybody what they wanted. I was the archetypal outsider who misread the situation I had been sent into. If I had lived in one of the neighbouring streets, I might have known that my timber castle was a bad idea.

The unhappy fate of my play-fort illustrates one of the strongest arguments for designing not just with people in mind but actually with people—in other words, in a collaborative or participatory way. Talking to people is a way to access local knowledge—of the twilight habits of disaffected teenagers, for example—which perhaps cannot be obtained by other means. It is better still if you can involve the community, including its estranged and marginalized elements, in the design process, so that when the park is laid out or the climbing frame set into position, the community sees it as its own, not as some random imposition from a distant authority. The skills required for this sort of work, which include listening, suggesting, explaining, negotiating, brokering, arbitrating, and a whole lot more, are seldom taught in the design studio. Also, working patiently with people can take a very long time because it is an iterative process, involving lots of meetings, lots of feedback, and lots of changes to the drawings. For designers who have been trained in the studio system to produce final designs by a given deadline, it can be hard to shift to a more patient, incremental progression. Yet unless this effort is made, much design, particularly in disadvantaged inner-city districts suffering high levels of delinquency, the time and money spent on new facilities can be wasted. Although it is possible to beef up specifications in an attempt to make items like park benches or play equipment more resistant to vandalism, this defensive approach has its limitations. It is far better to encourage local groups to take pride in their surroundings, because a sense of ownership can lead to forms of self-policing within communities. How best, then, to create that pride of ownership?

Participation

Back in 1969, an American planner called Sherry Arnstein published a paper entitled 'A Ladder of Citizen Participation' which became a classic text. She noted that there were degrees of participation which could be arranged like the rungs of a ladder. At the bottom was 'manipulation' which was really a parody of participation where officials would pay lip-service to local democracy by inviting a few well-chosen community representatives to join committees. It was the officials who 'educated' and persuaded the citizens, not the other way around. Only slightly better was 'informing', where the authorities had the decency to tell people about plans that could affect them, but again the flow of information was one way. Informing is certainly a necessary first step towards participation, but unless there are mechanisms for responding to feedback, it is no more than that. 'Consultation' involved soliciting citizens' opinions through such devices as attitude surveys, neighbourhood meetings, and public hearings. Arnstein cautioned that this could often be a sham, mere window-dressing, unless combined with more active modes of participation. Towards the top of the ladder were 'partnership', involving genuine sharing of power between officials and community representatives on policy boards and planning committees, and 'delegated power' where negotiations between citizens and public officials resulted in 'citizens achieving dominant decision-making authority over a particular plan or program.' Arnstein placed 'citizen control' on the very top rung, which ceded full authority for management and decision making to community insiders. Where public money is involved, those in power generally see full citizen control as too risky, and they may have a point. It is reasonable to ask whether the community has the skills and capacities required to control budgets and manage complicated operations on site. It is more usual, therefore, to find participation occurring as some form of partnership between local government and community groups, and this model is often successful.

Methods

Much has been written about community participation in planning and design, and there are literally hundreds of different approaches and methods. Many have been devised in response to the perceived failings of traditional methods such as questionnaire surveys and public meetings, which do not involve and empower citizens. They range from techniques for discovering what people want, such as briefing workshops, future search events, and guided visualization, to methods which harness local creativity, such as art workshops, model-making, and parish mapping, through to events and activities which engage citizens in the design and decision-making process itself. Corresponding to the top rung of Arnstein's ladder we have local development trusts, which are non-profit organizations, set up, owned, and led by a community to pursue social, environmental, and economic regeneration, usually working in conjunction with other private, public, or voluntary organizations. Perhaps more relevant to landscape architecture are a cluster of methods which bring together members of communities (insiders) with designers and others with specialist knowledge and skills, such as ecologists and engineers.

Charrettes and task forces. The term 'charrette' derives from the French term for a cart or a chariot. In the 19th century, architecture students at the École des Beaux-Arts in Paris would work furiously up to a deadline, upon which a cart would be pushed through their studios and they had to throw their designs into it for review by their teachers. The idea of intensely working against a deadline to solve a problem or produce a plan is captured in contemporary usage, which also suggests the idea of a focused team effort. Charrettes are short-term events, usually held over a few days, in which the participating professionals listen to the local stakeholders (not just residents, but other interested parties, such as politicians, sponsors, local business people, etc.), try to develop a collective vision for the place and then draw like crazy to meet the deadline. A task force or design assistance team is similar, in that it consists of outside practitioners brought into an area to work with the stakeholders to solve a problem, perhaps in response to some calamity like the closure of a major factory, a tornado strike, or a serious flood. At the heart of the process will be a multidisciplinary team of six to ten professionals who will work with the community for four or five intense, productive days. Some university landscape architecture programmes run charrette-like activities with local groups, often in conjunction with a studio design project, the products of which are made available to the community after the final review.

Workshops, Design Games, Planning for Real®. There are strong resemblances between all of these participatory methods. What distinguishes this group from the last is the emphasis placed upon people producing their own solutions rather than just contributing to solutions ultimately produced by outside experts. They generally involve some form of initial mapping or visualization stage. So in Planning for Real® (PFR), a process devised and trademarked some 30 years ago by Dr Tony Gibson, who was then part of Nottingham University's Education for Neighbourhood Change Unit, local people build a table-top model of their neighbourhood, which is then used in pre-advertised sessions held at various locations in the community, such as libraries or church halls. Participants place suggestion cards on the model showing the sorts of changes or additions they would like to see, such as a new park or play area, tree planting, better parking, or local shops. The cards can then be sorted and prioritized to produce an action plan for the community, to be followed up by working groups. Design games are similar. One version sometimes used for the planning of parks involves local people placing scaled cardboard shapes, representing facilities such as football pitches or tennis courts, or models of play equipment and park furniture, onto a plan of the park. These items have all been previously priced, so the participants can get a sense of what might be possible for a certain budget and can discuss and hopefully agree their own priorities.

I will close this chapter with two examples where participatory methods of planning and design have been used (see Boxes 1 and 2).

Box 1 The Ian Potter Foundation Children's Garden at the Royal Botanic Gardens Melbourne, Australia. Landscape architect, Andrew Laidlaw, completed 2004

A rusty metal gate featuring the shapes of old gardening implements invites children into this most magical of play gardens which was funded by the foundation created by Australian businessman and philanthropist Sir Ian Potter. The principal designer was Andrew Laidlaw, whose portfolio of work includes both renovation projects in botanic gardens and garden play spaces for schools. The garden, which won the 2005 Victoria Tourism Award for Best New Tourism Development, includes a long, curving, living tunnel, a snaking water rill, a secret ruin, and a rocky gorge surrounded by snow gums (*Eucalyptus pauciflora*) and tussock grasses (Figure 8). It was developed by a multi-disciplinary team which included expertise in visitor programmes, education, horticulture, and art, but it also involved working with children from two primary schools, one from the inner-city and one from a rural bush environment. The design team visited the schools to gather ideas for the garden from the children. On a subsequent occasion the designers presented the children with a concept plan which showed them some of the main elements, such as spirals, tunnels, and grassy mounds, then the children were invited to draw the features which they wanted the most. The children were also brought to the Botanic Garden and encouraged to interact playfully with the plants. An artist worked with them, creating artworks in free play. Through such activities, the designers learnt that the children's play was active and energetic, and that they enjoyed places where there was a sense of spatial enclosure. A review of the methods used observed that the interactive activities in the Gardens were much more helpful to the designers than the interviews with the children at their schools at the concept stage. The project is exemplary in the way it sought to develop the garden with children, rather than just for them, and for its adoption of active, creative, and enjoyable methods that would appeal to the children as creator-users of the space. Though this garden was made in the high profile setting of a major botanic garden, the same philosophy can be applied to the improvement of school grounds or the creation of community parks and play areas.

8. The Ian Potter Foundation Children's Garden at the Royal Botanic Gardens Melbourne, Australia (completed 2004)

Box 2 Groundwork UK

The charity Groundwork is the largest single employer of landscape architects in the United Kingdom. Founded in 1982, Groundwork is an organization which mobilizes local people and resources to improve the prospects of struggling communities, using the local environment as the focus for action. The idea took root in places suffering from the decline of traditional industries such as coal, steel, and quarrying, where communities had not only lost their main sources of employment, but also had to contend with the ravaged landscapes which industry had left behind. Although good design was always an objective, the initiative was also about resolving social tensions, improving people's life chances through training, education, and work experience, attracting investment, and stimulating the local economy. The organization also helps people to think about the ways in which they can take local action to counter global environmental problems. Landscape architects work alongside community development officers, youth workers, and project officers, engaging patiently and enthusiastically with individuals and community groups to harness creativity and bring about beneficial change. Groundwork is now a federation of around 30 local non-profit trusts, which between them deliver thousands of projects every year. It almost seems invidious to select any one project or even region, but to give some sense of Groundwork's activities we can consider Groundwork Leeds, which in a single year helped young people to redesign and renovate a skate-park, worked with a local residents' association to manage an overgrown woodland for public access, redesigned a dilapidated playground as an informal playground incorporating natural elements, and renovated Victoria Gardens, a prominent open space in the city centre, with input from the children of Little London Primary School. Many Groundwork projects are small scale and not many make the pages of glossy design magazines, but that is hardly their aim. It is the cumulative impact over three decades of Groundwork's many projects upon people's lives that ought to be considered. The approach has been so successful that it has inspired the creation of sister organizations in the USA and in Japan.

Chapter 8
Making good again

Back in 1989, a prominent campaigner against the motor car wrote a letter to the professional magazine then read by most British landscape architects and students. His purpose was to complain about landscape architects working on road proposals, but he smeared the whole discipline when he described landscape architecture as 'the Nightsoil Profession' and said that it specialized in clearing up the messes made by others, rather than preventing them from being created in the first place. Living and working for most of my life in the north-east of England, never far from the coaly river Tyne, I have seen a lot of the mess that industry can make, but I think the task of clearing it away is a noble one. When the environmental artist Mierle Laderman Ukeles (1939–) began her long spell as artist-in-residence with the New York City Sanitation Department, one of her first public acts was to shake the hand of every refuse collector and thank them for the vital work they did for society. This sort of work is often invisible and poorly remunerated. Ukeles was suggesting it should be re-evaluated. Landscape architects and reclamation engineers deserve similar thanks. When the writer J. B. Priestley visited the north-east of England in 1933, which was then in the depths of an economic depression, he wrote 'I never saw a bit of country that was in more urgent need of tidying up.' A visitor today would have difficulty in spotting a single coal tip, while large stretches of the banks of the once industrial Tyne are verdant with grass and trees.

Landscape architects have played an important, though often unsung, role in this transformation, but this is not a merely local phenomenon; they are doing the same for post-industrial landscapes all around the world.

What's in a metaphor?

Land reclamation is virtuous work, and many landscape architects of my acquaintance say that it is this aspect of their occupation that has given them the most satisfaction, but this does not exempt it from criticism. One line of attack is to say that it is merely a cosmetic exercise, like sweeping dirt under the carpet. This has some force, because when dealing with dirty sites one of the problems is the disposal of toxins. If soil is contaminated, it makes little sense to take it elsewhere; it needs to be dealt with on the site. If nothing can be done to reduce toxicity, the remedy is to push it to a remote part of the site, to encase it in an impermeable clay covering, then to spread soil and sow grass over the top. This does, indeed, resemble a poor housekeeping practice and it often means that the part of the site which becomes the 'hot spot' can never be built over or dug into. Nevertheless, it was often the best available solution, given the technologies that were available. It may be that developments in phytoremediation (the use of plants to neutralize toxins) and nanotechnology can provide effective and permanent ways of dealing with site contamination. Ironically, Geoffrey Jellicoe's dismissal of 'seemliness' as a sufficient objective (contained in his speech to the Institute of Contemporary Arts in London in 1961, which perhaps accounts for its thrust) was made at just about the time that the major programmes of land reclamation got into their stride. The decline of heavy industry and manufacturing in many Western countries has ensured a steady stream of commissions. In times of economic turmoil, derelict sites are often created more quickly than they can be reclaimed. This work is necessary and it seems injudicious to denigrate it as mere tidying up or the pursuit of meagre seemliness. Contemporary landscape architecture is, in many ways, opposed to the merely scenographic. Other metaphors might be used. It is commonplace now to talk about 'recycling' derelict (brownfield) land, linking the practice to good environmental management. The re-use of brownfield sites, for housing for example, can provide alternatives to development on farmland which contributes to urban sprawl. Or we could employ the language of healing, with the landscape architects and engineers cast as the surgical team called in to treat the wounds and scars inflicted on the landscape by industry. We might also invoke the image of the fine art restorer, repairing the damage done over years to an old masterpiece. This is certainly one way of seeing the applied science of restoration ecology, which is often brought into reclamation projects, the aim being to recreate the sorts of habitats that would have existed prior to disruption by industry.

Boxes 3, 4, and 5 provide examples of this kind of work.

Box 3 Turning the tide: the Durham coast, UK (1997–2003)

For 150 years coal waste was dumped on the beaches of the Durham coast. At its zenith the coal industry tipped 2.5 million tonnes of waste each year, amounting to 40 million tonnes of waste over its period of operation. The county's infamous black beaches provided bleak backdrops for the films *Get Carter* and *Alien* 3, but when the pit at Easington closed in 1993 it marked the end of a grimy era and the way was at last clear to tackle the appalling environmental damage that mining had caused.

In 1997 the £10 million Turning the Tide project, a partnership of 14 organizations, began to grapple with what must have seemed a Herculean task. Two large spoil heaps at Easington and Horden were removed, to prevent the material they contained from being washed out by the tides to become a pollution hazard on the nearby beaches. The mechanical apparatus and concrete towers were also demolished. Coastal footpaths were improved, cycle paths created, and new limestone grassland established on cliff tops and headlands. The aim was to recreate the sort of landscape character that would have existed prior to the industrialization of the coast. In this the project was so successful that it was recognized as UK Landscape of the Year in November 2010 and runner up in the Council of Europe Landscape Award 2011.

Box 4 Bundesgardenshauen, garden festivals, and expos

There is a category of events which recognizes that the reclamation of derelict land is a cause for celebration. Germany led the way in this by reviving the tradition of the national garden show in 1951. The Bundesgartenshau was held every two years in a different city and this became a mechanism for treating war-damaged sites. After a Bundesgartenshau, the exhibition landscape would be converted into a permanent piece of public parkland. The first of these events was held in Hannover and, at the time of writing, the programme is set to continue with shows in the Havel region in 2015, the old Tempelhof Airfield in Berlin in 2017, and in Heilbronn in 2019.

The British government experimented with a version of the Bundesgartenshau between 1984 and 1992; the first National Garden Festival was created on a former dockland site in Liverpool, and others followed at two-year intervals in Stoke-on-Trent, Glasgow, and Gateshead, with the last taking place on the site of a former steelworks in Ebbw Vale in Wales in 1992. The aim was not just to create parkland but to attract inward investment. The landscape design was often compromised by political interference and confusion about objectives, but the festivals attracted millions of visitors and accelerated the pace of land reclamation, if not always economic regeneration, in places which needed a boost.

A similar pattern is found at Expos and other major international events. For example, the site of Expo 2010 Shanghai was held on both banks of the Huangpu River and included a 14 hectare site previously occupied by a steelworks and a shipyard. The design went beyond a mere clean-up. By incorporating a constructed wetland and ecological flood control measures, the designers from the Chinese office Turenscape were able to use plants to absorb pollutants from river water and this water was then used throughout the Expo for non-potable purposes.

Box 5 Crissy Field, San Francisco, USA (1997–2001)

This is different from the sites mentioned so far, in that its former use was military rather than industrial. It had been an airfield constructed upon a tidal marshland on the northern waterfront of San Francisco's Presidio, a 647 hectare military complex abandoned in 1994. Its soils and groundwater were seriously contaminated by aviation fuels, pesticides, and solvents used for cleaning aircraft.

The initial clean-up, undertaken by the army, involved the excavation of severely contaminated soils which were incinerated off site and replaced with native soils from elsewhere in the Presidio. Less contaminated soils were heated in a mobile kiln in a process called 'low temperature thermal desorbtion', which extracted organic contaminants and left the resulting dust clean enough to be interred on site.

Landscape architects Hargreaves Associates were commissioned by the National Parks Service to provide a plan which responded to the natural and cultural history of the place. Sculptural landforms which mimic and amplify the effects of wind and wave action were created on an otherwise flat site. The design included the restoration of tidal wetlands where birdwatchers have subsequently sighted 135 species. There is a beach, which is popular with windsurfers, and the site provides stunning views of Golden Gate Bridge.

Manufactured sites

Professor Niall Kirkwood of Harvard Graduate School of Design used the term 'manufactured sites' as the title for a book about approaches to land reclamation. The title was a kind of pun; not only were the sort of sites he described found in older manufacturing districts, but the sites themselves had been manufactured as accidental by-products of industrial activity. They owed their present character to this history. What is more, to turn them into something of benefit to society, they had to be remade. In some cases the very materials used to do this, for example clean soils, would have to be manufactured on the site. Reclamation is typically the field in which two different sorts of expert must work together. On one side are the site designers (landscape architects, planners, urban designers) while on the other are the civil and environmental engineers. We might expand this second group to include environmental scientists and ecologists. No single specialist can solve all the problems involved in bringing a contaminated brownfield site back into beneficial use. Landscape architects have often demonstrated that as generalists and synthesizers they are the most effective orchestrators of this collaborative enterprise.

The problems encountered on such sites can be daunting. Toxicity is the most troubling but may be the least obvious. Contaminants such as heavy metals, oils, and chemical residues can be invisible to the naked eye. Workers on the most seriously contaminated sites must wear protective clothing to prevent skin contact with toxic substances. Sometimes there is value in the materials found on the site; for example a technique known as coal-washing can be used to recover saleable coal from mine wastes, often contributing significantly towards the costs of the clean-up. Mechanical ways of dealing with contaminated soils are being supplemented by biological techniques. Some plants, referred to as phreatophytes, take up large quantities of water and have been found useful in treating groundwater pollution. A study at a former gasoline transfer terminal at Ogden, Utah, for example, showed that poplar trees could inhibit the flow of groundwater and enhance petroleum degradation, effectively preventing contaminants from leaving the site. Plants known as hyperaccumulators (Indian mustard and sunflowers are examples from the United States) have been used to absorb heavy metal pollution from toxic sites. It is sometimes even possible to harvest the plants and reclaim the minerals. Phytoremediation would seem to be one of the most promising techniques for reclaiming land, and, at face value, it seems gentle and environmentally friendly, but much research remains to be done. For example, if the contaminants are not harvested, what are the ecological consequences of the contaminated plants entering the food chain? These matters are complex and require multi-disciplinary investigation.

Completed landfills are a category of site which often requires the attention of a landscape architect. Landfills are the places where society entombs its refuse, often encasing it in an impermeable casing of clay or plastic. Landfills which include organic matter produce the greenhouse gasses carbon dioxide and methane, and the latter is highly flammable. Housing development is not usually permitted on gas-producing landfills, not simply because of the risk of fire from the methane but because of the risk of subsidence from the settlement of materials within the dump. As a result, many landfills are recycled as public open space, but they bring with them particular problems. It used to be thought, for example, that tree planting over clay-capped landfills was a problem because tree roots might penetrate the seal, although more recent research suggests that this is not such an issue. Leaking methane gas can, however, impede the growth of trees.

Despite these obstacles, there is a tradition of landfills being turned into parks. The first in the United States was the ironically named Mount Trashmore in Virginia Beach, VA (opened 1973), a 68 foot (20.7 metre) artificial hill, constructed by sandwiching layers of clean rubbish between layers of soil, producing a place which is still popular with families and kite-flyers. Byxbee Park, created in the late 1980s on a small section of the Palo Alto city dump in California, is another celebrated example. Landscape architects Hargreaves Associates collaborated with the artists Peter Oppenheimer and Peter Richards to create a landscape which, in subtle ways, acknowledges the 60 feet of refuse upon which it is built. The windy, coastal park, not far from the city airport, has no trees because of the anxiety over root penetration. The flare which burns off excess methane from the tip was incorporated into the design, which also includes a forest of half buried telegraph poles. These were planted straight, but over time they will tilt as the garbage settles beneath them. A chevron earthwork is a sort of visual joke. On an aeronautical map chevrons mean 'don't land here'—and aeroplanes don't, but in winter large numbers of geese settle briefly on their way to southern warmth.

Keeping the heritage

Critics of land reclamation sometimes point out that it can destroy local heritage. Writing about the erasure of traces of coal mining from the Dearne Valley in South Yorkshire, the ecologist and priest John Rodwell wrote that in the case of a human individual we would 'regard memory-loss as a pathology worthy of concern'. People can be uprooted, even if they do not move. If the rapid industrialization of land is regarded as a trauma, and the sudden closure of industries that may have sustained communities for generations is another, then perhaps hasty reclamation is yet another. It does not have to be, however. A project in the 1970s by American landscape architect Richard Haag (1923–) pioneered a different approach. He convinced the city of Seattle, which was setting about the reclamation of a former gasworks on the north shore of Lake Union that the rusting relics of the old gasification plant did not have to be removed. Instead these imposing pieces of industrial history were retained as a central feature of what became Gas Works Park. His example was not much copied until the German Latz + Partner began a similar practice in the

9. In the 1990s the German landscape architecture office Latz + Partner turned a former steelworks in the Ruhr valley into the much celebrated Landschaftspark Duisburg Nord

post-industrial landscapes of Saarbrucken and the Ruhr Valley. Their best known work is the former steelworks now known as Landschaftspark Duisburg Nord, where the gently corroding furnaces have been kept, gardens have been planted in old ore bunkers and a diving club makes use of a disused gasometer, where an artificial reef and a wrecked motor yacht now add interest to the underwater landscape (Figure 9).

In Germany, this recycling of sites has almost become the mainstream approach and there are numerous examples of landscape architects conserving rather than erasing industrial heritage. Planergruppe Oberhausen, another landscape architecture practice active in the Ruhr-Emscher region, has been working on the Zollverein coal mine and coking plant. The mine, with buildings designed by Modern Movement architects, was known as 'the most beautiful colliery in the world'. After closure in 1986, it was threatened with demolition, but a campaign led to its listing as a World Heritage Site by UNESCO in 2001, and 50 million Euros of public money were committed to its transformation. The eminent Dutch architect Rem Koolhaas was engaged to masterplan the site and to convert the former coal-washing building into the Ruhrland Museum. Agence Ter, a French landscape practice, was also involved in developing the open space design. Landscape architects, working in collaboration with other designers and artists, and in a participatory way with local schoolchildren, have focused upon the park, designed the paths and cycle-ways in the park around existing features including the harp-shaped system of railway sidings. The gigantic industrial complex remains, but it is softened by the groves of birches and willows that have colonized the spaces between the buildings. The designers have developed a promenade around the former industrial buildings with solar-powered accent lighting to enliven the scene in the evenings and the artist Ulrich Rückriem, celebrated for his monumental granite pieces, has established a sculpture park within the post-industrial forest.

Chapter 9
Landscape planning

Many years ago, delegates at a European conference of landscape architecture academics were presented with a task. They were each given a list of terms—'landscape architecture', 'landscape design', 'landscape planning', 'garden design', 'garden art', and so on—each represented by an arbitrary geometrical shape, and they were each asked to produce a drawing which showed the correct conceptual relationship between these fields. So, if garden design was represented by a square and landscape architecture by a circle and a delegate thought that garden design was completely encompassed by landscape architecture, he should draw a circle enclosing a square. If two fields might be said to overlap, they could be represented by two shapes that crossed. To no one's great surprise, every diagram that emerged was different.

It might be interesting to repeat this experiment, after more than two decades of international discussion and progress towards integration, but I expect that even now there would not be total agreement. I hope there would be general recognition that the overall title of the discipline is 'landscape architecture', despite the shortcomings of that name. I also think there would be recognition that under this umbrella term there are two complementary activities called 'landscape design' and 'landscape planning'. These undoubtedly overlap, but I am going to try to distinguish between them. The conventional way to do this is to draw up a list of binary oppositions, as follows: design vs. planning; artistic vs. scientific; small sites vs. extensive regions; creative vs. problem-solving; synthetic vs. analytic; serves individuals vs. serves society. This list gives some sense of the difference between two sorts of activity. It is also undoubtedly true that some landscape students are instinctively drawn toward the more artistic aspects of landscape architecture, while others are happiest when analysing survey material, preparing plans, or writing reports. However the inadequacy of this binary split was demonstrated by Professor Richard Stiles from the Technical University of Vienna, who observed that landscape cannot easily be split into small sites and large territories. It is a continuum, with intimate, garden-like spaces at one end, neighbourhoods and networks in the middle, and regional landscapes and whole watersheds at the other end. Similarly, it is impossible to say that creative design involves no analytical thinking or problem-solving. It generally does. Stiles was pointing out that planning and design were not really separate activities with distinct theoretical bases. The relationship between design and planning is more like the yin and yang symbol: there is always a bit of planning in design and a bit of design within planning (this is my observation, not Stiles'). There are, however, paradigm cases. When a landscape architect prepares a design for a site which is entirely within the control of a private client, such as a garden, the activity would usually be called design, though it includes aspects of planning, such as working out where the best place for the vegetable plot might be. A landscape architect often has to employ much the same skills when designing a park or a public square, though the client is no longer a private individual. Managing large estates on behalf of large landowners, whether individual or corporate, is often a rationally based activity, but aesthetics may nevertheless come into consideration—it would be usual, though, to think of this as planning. Finally, we have the classic landscape planning scenario, where the landscape architect must prepare a plan on behalf of a local administration for the land over which it has jurisdiction. Again, this might seem to be a largely rational decision-making process, but it often has to take aesthetic, cultural, and even spiritual values into account.

Origins of landscape planning

Landscape planning's origins lie in the anti-urban attitudes fostered by Romanticism and Transcendentalism, and from an urge to protect nature against perceived human encroachment. As we saw in Chapter 5, Frederick Law Olmsted, the father of landscape architecture, worked with naturalists to secure the protection of what they thought of as pristine landscapes in the American West. The philosophy underlying this is aptly summed up in Thoreau's aphorism 'in Wildness is the preservation of the World'. For Thoreau and for Olmsted 'wilderness' and 'the West' were synonymous, so preservation meant the exclusion of human activity. As the environmental historian William Cronon has pointed out, this conception of wilderness was flawed from the outset, because it ignored the fact that these so-called wildernesses had for centuries been the cultural landscapes of Native Americans. Nevertheless it gave us the notion of the designated and protected landscape and, specifically, of the national park. In America this meant a place where no people lived at all, but when Britain designated its first national parks, shortly after the Second World War, these were cultural landscapes such as the Peak District and the Lake District, where the character and aesthetics of the landscape depended on centuries-old farming practices. The idea of 'natural beauty' is enshrined in much British planning legislation, although it is a very slippery concept to define. Countryside is, after all, a hybrid of natural and human processes. Britain has a swarm of landscape designations, some related to history, some to scarcity of habitat, some to cultural significance and scenery, and all bearing upon the drafting of development plans, thus helping to determine what may be built where.

There are international designations too, such as the Ramsar Convention which lists globally significant wetlands, recognizing their scientific, ecological, and cultural value; but soaring above all of these, at least in status, is UNESCO's listing of World Heritage Sites. To get into this elite category a site must have 'outstanding universal value'. It was established with the idea of protecting our collective global patrimony, safeguarding cultural gems like the Acropolis or natural wonders such as the Grand Canyon and the Great Barrier Reef, but the rules were later changed so that cultural landscapes, hybrids of culture and nature, could also be included. Some designed landscapes, such as the classical Chinese gardens of Suzhou are included as cultural artefacts, but so is the Göreme valley in Cappadocia, Turkey, where vernacular homes were carved out of the soft rock amid a spectacular landscape of natural pinnacles and towers.

As we have already seen in earlier chapters, this desire to protect the perceived beauties of nature was balanced by a mission to improve living conditions in overcrowded cities. The benefits of open space within cities had long been recognized and promoted. William Penn, the founder of Philadelphia prefigured the Garden City Movement with his vision of a 'greene Country Towne'. He had survived London's bubonic plague in 1665 and the great fire of 1666, so he wanted buildings to be set in large plots surrounded by open space so that the new city 'will never be burnt and will always be wholesome'. Notice that Penn's plan promoted both safety and public health; the idea that open spaces can provide benefits of many different kinds is still at the heart of contemporary thinking—in contemporary jargon it is 'multifunctional' and 'cuts across multiple policy agendas'. Planning-speak may be opaque, but at least it is colourful: this totality of parks and gardens can be considered 'greenspace', never to be confused with 'brownfields', but there is also 'bluespace', the collective term for rivers, lakes, ponds, and other water-bodies within the urban fabric. Olmsted's Emerald Necklace, a linked chain of parks in Boston and Brookline, MA, would nowadays be described as a 'greenspace network' which is not quite so poetic, but essentially the same idea. Patrick Abercrombie's Greater London Plan of 1944 was based on a systematic survey of the existing landscape and recommended the establishment of a Green Belt to contain sprawling urban development and a system of open spaces based around parks, green spaces, and river corridors.

McHargian landscape planning—landscape suitability

Ian McHarg, too, was concerned about unrestricted development. The idea that some developments are more suitable to some landscapes than to others seems only commonsense, but when homes built on floodplains are inundated or hotels built on cliff tops fall into the sea, the extent of human folly becomes evident. We could avoid such calamities and live more harmoniously with nature, thought McHarg, if we took natural processes and values into account. He proposed a method for bringing everything into the picture. Known as 'landscape suitability analysis' or sometimes just as 'sieve-mapping', the technique he developed involved layering information on acetate sheets. So, for example, in considering the optimal route for a new highway, McHarg would combine layers showing the engineering properties of the substrates with layers showing productive soils, significant wildlife habitats, important cultural sites, and so on. When these were combined, it was the areas which were clearest of symbols that were the better areas in which to construct the road. The method also worked well for planning development at a regional scale. Typically, after gathering physiographic, climatic, and geological data, McHarg could produce suitability maps, usually zoned for agriculture, forestry, recreation, and urban development. The method, which relied on extensive gathering and manipulation of data, became much easier with the growing availability of computers, and 'McHarg's Method' became the basis of the technology known as GIS (Geographical Information System) which uses digital map layers instead of superimposed drawings. The advent of landscape ecology also enriched McHargian landscape planning. It provided the theory to explain why some ecosystems might decline and also suggested principles whereby they might be safeguarded and improved.

In general landscape planning does not begin with a blank slate. The polder landscapes created in the Netherlands in the 20th century are the exception that proves the rule. Here, where new land was won from the sea, it was possible to start from scratch, planning farms, dikes, road, settlements, and woodlands. These flat landscapes, with their straight lines and rectilinear shapes are the epitome of rational planning, but they have their own striking aesthetic. Most places, however, do not come off a drawing board in quite the same way. Most landscapes have developed over centuries and they are layered. The term 'palimpsest' (the name for a Roman tablet or mediaeval scroll, partially erased to be re-inscribed) is often used to express the idea that even if a landscape is altered, traces of its history will still remain. Most landscape planning begins with something rather complicated, and we cannot even say that this is a complex object because, as many theorists have pointed out, landscape is something mental as well as physical, something subjective as well as objective.

From special landscapes to the whole landscape

Though the origins of landscape planning were in the designation and protection of areas of countryside considered to be exceptional, the adoption of the European Landscape Convention in 2000 (a Council of Europe treaty) signalled a significant shift in thinking. The Convention's definition of landscape as 'an area, as perceived by people, whose character is the result of the action and interaction of natural and/or human factors' recognized that landscape was not just physical. It was something 'perceived by people', in other words, something both understood and shared. Landscape was recognized as 'an essential component of people's surroundings, an expression of the diversity of their shared cultural and natural heritage, and a foundation of their identity'. In the Article headed 'Scope' the Convention states that it 'applies to the entire territory of the Parties and covers natural, rural, urban and peri-urban areas. It includes land, inland water and marine areas. It concerns landscapes that might be considered outstanding as well as everyday or degraded landscapes.' This does not mean that the old protective designations around such places as the Plitvice Lakes National Park in Croatia or the Pyrenees National Park in France are about to disappear, but it does mean than politicians and planners have to also consider policies for recognizing and conserving the qualities of everyday landscapes, closer to where most people live, and improving landscapes which are considered to be failing in social, economic, ecological, or aesthetic terms.

The European Landscape Convention also marks a big shift away from decision-making by experts towards decision-making by ordinary people. What it actually calls upon signatory states to do is to 'establish procedures for the participation of the general public' alongside local and regional authorities in the definition and implementation of landscape policies. It is all open to interpretation, of course, and doubtless the practice in various countries will differ, but it is a shift nonetheless. Landscape planners will no longer be able to rely upon their own judgements, no matter how well informed they may consider themselves to be. Experts will still be needed, but expertise in facilitating participation will be at a premium. A campaign is currently pressing for an International Landscape Convention, backed by the United Nations, so these considerations may soon apply to the whole world.

Assessing the quality of any landscape is a question fraught with difficulties. Attempts to do this quantitatively, by scoring the features found in map squares, for example, were eventually abandoned and replaced, at least in England and Scotland, by a method known as 'landscape character assessment' (LCA), developed in the 1980s by the office Land Use Consultants, which endeavours to separate the description of landscapes from any judgements that might be made about them. A complementary method, known as 'historic landscape characterization' (HLC) adds 'time-depth' to the description. In line with the European Landscape Convention's shift away from red-line designation of special landscapes, HLC is particularly concerned with how to protect and manage dynamically changing rural landscapes. If we admire landscapes because they are palimpsests of past change, logically we must be prepared to allow further change. The question is what scale and speed of change is acceptable, and here, once again, it is important to engage with public opinion.

Environmental impact assessment and visual impact assessment

Much of the planning work which landscape architects do is tied to particular development proposals. They may be working on behalf of a developer in the quest for planning permission required before a project can go ahead, but they might also be working on behalf of a local authority, assessing a proposal once it has been submitted, or on the behalf of objectors trying to show that a particular development will be harmful. The range of projects can vary from something relatively small scale, like a small housing development in a field outside a village, to something very large, such as a new airport or high-speed railway line. Many countries have adopted a procedure called 'environmental impact assessment' (EIA) which obliges the promoters of particular classes of development to undertake a comprehensive review of any likely effects of the proposal and any steps that might be taken to mitigate them. The sorts of project covered by the European legislation on EIA includes such things as oil refineries, motorways, chemical works, open-cast mines, waste disposal sites, and quarries, but the list also includes large, intensive poultry farms.

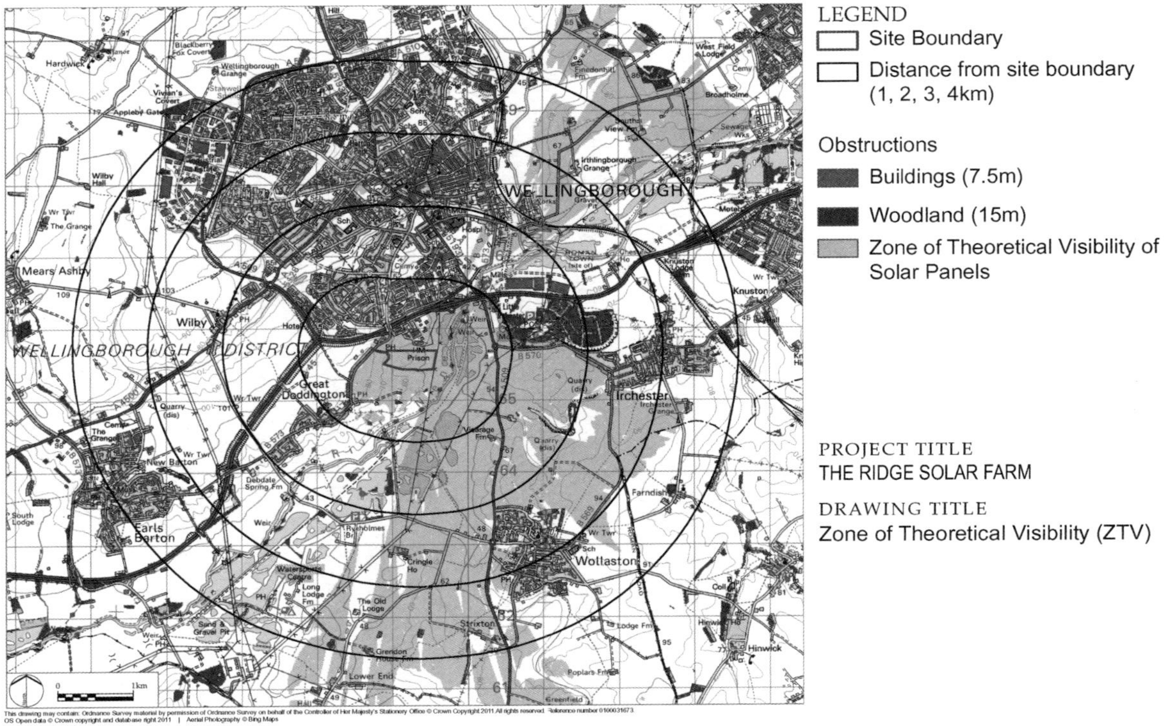

10. Zone of Theoretical Visibility plot for the Ridge Solar Farm near Wellingborough, prepared by Landscape Design Associates, 2013

EIA includes a separate but linked procedure called 'landscape and visual impact assessment' (LVIA) which looks at the possible effects of the development upon the physical landscape and upon views and visual amenity. LVIA is generally undertaken by landscape architects. It can, of course, be done for projects which do not require a full EIA. The virtue of carrying out these assessments early in the development process is that they can serve as design tools, identifying ways to avoid impacts or to reduce their scale. When assessing the visual impact of something like a wind farm or a new factory, the landscape architect used to go and stand on the site of the proposed structure and with map in one hand and pencil in the other, and try to map the area from which it could be seen. This would be supplemented by drawing sections based on map contours (Figure 10). There are now computer programs which enable this 'zone of theoretical visibility' (ZTV) to be estimated much more accurately. Visualization software can also provide reliable images of what a proposed development might look like from particular viewpoints. Problems can sometimes be avoided by reducing the scale of the development or by some adjustment of levels. Experience has shown that such measures are often more effective in mitigating impact than cosmetic landscape works such as the planting of trees to screen something deemed unsightly.

Green infrastructure planning

Urban green space tends to get taken for granted. People like it and sometimes pay a lot of money to live near to it, yet its maintenance is often one of the first things to be cut in times of austerity, and it tends to get nibbled away for both public and private projects. Every so often the case for green space needs to be articulated again, and the way in which the argument is presented usually reflects the preoccupations of the times. We live in an era when arguments made by economists hold enormous sway over public policy. If green space is perceived as little more than ornament, hard-headed accountants are likely to conclude that its upkeep is too much of a drain on the public purse. Hence the need for arguments which show that green space is useful and functional, that it 'provides cross-cutting benefits across policy agendas', and that it does something for us. The latest form this advocacy has taken is green infrastructure planning.

Green infrastructure planning builds upon the idea of 'environmental services' mentioned in Chapter 5. Many of the historical examples of park provision we have already considered could easily be reclassified in these terms. Olmsted said that Central Park would be 'the lungs of the city' and Boston's Emerald Necklace, constructed from 1878 onwards, can be regarded as a successful green infrastructure project which delivered public health benefits from improved sewage treatment. Contemporary thinking classifies ecosystem services under a series of headings. There are 'supporting services' such as the creation of soils, which underpin all other services. There are 'provisioning services'—the supply of necessities like food and fuel. There are 'regulating services' including the capture of carbon from the atmosphere. Finally, there are 'cultural services' which include all the ways in which nature contributes to human well-being. Here landscape plays a significant role, contributing aesthetic inspiration and enjoyment, a sense of history and place, recreational opportunities and spiritual elevation. Well planned green infrastructure can bring these benefits to people in densely populated urban areas. It can help to alleviate some of the problems caused by climate change: green spaces can, for example, be designed to retain large volumes of flood water and to help it to percolate into the ground, thus safeguarding built-up areas. A crucial idea is 'multi-functionality', the concept that many different functions or activities, ranging from water management, to habitat protection, to the provision of health enhancing outdoor recreation, can be provided by the same pieces of land.

Chapter 10
Landscape and urbanism

Though the word 'landscape' is often taken to be synonymous with 'countryside' or with 'rural scenery', the examples of landscape architectural work given throughout this book have shown that landscape architecture is an urban practice as much as a rural one. Indeed, I would go further and say that it is much more about what happens in and around towns and cities than it is about farmland or bucolic beauty spots. Landscape work is usually tied in some way to development, and most of this happens in urban areas. What is more, in the future most people are going to live in cities. According to the United Nations the percentage of the world's population living in urban areas has already overtaken the percentage living in rural areas. This trend seems set to continue, so that by 2050 it is projected that 70 per cent of people will live in towns and cities. More and more people will live in 'megacities' which are defined as urban agglomerations with a population of over 10 million. Tokyo is currently the largest city in the world with a population exceeding 34 million. There were 16 megacities in 2000, but it is estimated that there will be 27 by 2025 and 21 of these will be in less developed countries. The rapid and largely unplanned expansion of cities during the Industrial Revolution was the spur to improvements in sanitation, housing standards, and the provision of urban parks. In the same way, the current spate of urbanization poses questions about the possible quality of life in such vast settlements, challenging landscape architecture and other disciplines concerned with the built environment to adjust to the scale of the megacity phenomenon.

Landscape architecture and urban design

The task of shaping liveable cities is so complex that it calls for contributions from a range of professionals, including landscape architects, building architects, and urban planners. Each of these disciplines has its own sensibilities, its own approach to training and education, and its own specialist knowledge. A landscape architect, for example, should know which species are the best street trees and how to plant them in sidewalks stuffed with sewers, gas mains, and fibre-optic cables, but an urban planner might have a better grasp of residential densities or the relationship between travel patterns and built form. One of the strengths which landscape architecture characteristically brings to urban issues is an advanced understanding of natural systems and ecology. In the late 1950s there was already a growing conviction that the problems presented by cities, particularly in the age of the automobile, required a pooling of expertise—and a series of conferences held at Harvard University attempted to find a common basis for a new discipline called urban design. This discipline is now well established in its own right and there are numerous university programmes, generally at post-graduate level, to prepare practitioners to work in this field. Yet, unlike landscape architecture or planning, urban design is not a profession with formal accreditation procedures and the accompanying institutional paraphernalia. Although this can be liberating, it also means that to practice as an urban designer one generally also needs to be qualified in one of the related professions—perhaps landscape architecture.

Unsurprisingly, considering how urban design came into being, there are significant overlaps between urban design and landscape architecture. We can return again to Olmsted's practice, a lot of which might now be classified as urban design as much as it was landscape architecture. Planning a city neighbourhood involves designating and designing parks and open spaces, as well as housing areas, shopping centres, and transport systems. Olmsted's park systems, as we have seen, solved problems of urban sanitation as well as providing places for recreation. Designing a park necessarily involves an understanding of its context, its position in the urban fabric, its relationship to the places where people live and work, its connections to streets as well as to other open spaces, and the ways in which residents and visitors to the city are likely to use it.

The differences between landscape architecture and urban design are largely a matter of perspective. This became clear to me when running joint studios for landscape architecture and urban design students. When presented with the same urban site, usually an existing open space or land reclaimed from derelict industry, the urban designers' inclination was to fill it with buildings, amongst which would be a scattering of small parks and urban squares. The landscape architecture students tended towards the opposite direction, scattering a few buildings amid large tracts of open space. The urban designers tended to see green open space as ornament and occasional relief, missing the functional and ecological benefits of larger parks and connected greenspace systems. The landscape architects, conversely, lacked confidence in dealing with built form, and at worst their designs did not seem to belong to a city. The purpose of such joint studios, of course, is to overcome these blinkered perceptions and to become aware of what other disciplines have to offer. Landscape architects need to be aware of the economics of urban development, for example, but need not be experts in them. Urban designers ought to understand the potential of green infrastructure, but can safely leave its implementation to landscape architects.

Suburbanization, sprawl, and many varieties of urbanism

There is a strong connection between transport systems and urban form. Back in the 1930s there was already an outcry in Britain about 'ribbon development' along trunk roads and the way this was joining towns together and harming the aesthetics of the countryside. Uncontrolled development was seen to be devouring pleasant scenery. In America, where land was plentiful and gasoline was inexpensive, the extension of large suburban zones around original downtown districts was labelled 'sprawl'. Despite the apparent consumer preference for and aggressive marketing of suburban lifestyles, sprawl was condemned by most architects, landscape architects, and planners. It has frequently been linked to a host of social and environmental ills. Suburbs are thought to lack the vibrancy and sociability of traditional neighbourhoods, while encouraging a dependency on the car which promotes unhealthy lifestyles and an epidemic of obesity. America is also the world's largest per capita emitter of carbon dioxide, something which has also been linked to low-density living and a love-affair with the automobile. The vision of spacious living promoted by Thomas Jefferson and William Penn has turned out to have serious repercussions.

There have been many responses to the issue of sprawl, many of which share salient features. The earliest was New Urbanism, an urban design movement which opposed the dispersion and atomization of communities by seeking to recreate a lively civic realm and a strong sense of place. It emerged in the 1980s drawing upon the urban visions of architect Leon Krier (1946–) and the 'pattern language' of the theorist Christopher Alexander (1936–), both of whom advocated a return to time-honoured ways of building cities derived from Europe. It advocated walkable neighbourhoods, often centred upon a park or an urban square. Narrow streets, some lined with trees, would discourage traffic and all the components of a liveable town—schools, nurseries, play areas, and shops—would be easily reached on foot. Stylistically the movement tended to be conservative and backwards-looking, seeking to replicate traditional styles of building. Two of the best known examples of New Urbanist inspired developments are Poundbury, on the outskirts of Dorchester, UK, and Seaside in Florida, but critics have found an ersatz quality in these places, which is perhaps why Seaside was selected as the location for *The Truman Show* (1998) a film in which the central character unwittingly lives a near-perfect, artificial life, manipulated by the production team of a television programme. Next there came ideas of 'smart growth', the 'compact city', and 'urban intensification', all of which retained New Urbanism's ideas about pedestrianized urban centres, without its nostalgic yearning for the 19th-century European city. Characteristic of such approaches are the provision of a range of housing choices, well integrated mass transport systems, mixed land uses, and the preservation of farmland, urban greenspace, and environmentally significant habitats. It is clear that landscape architects have a large role to play in the realization of such a vision. This way of thinking has been very influential in several European countries, particularly the UK and the Netherlands.

Transport and infrastructure

The trams in Barcelona, Strasbourg, and Frankfurt glide charmingly between avenues of trees, along ribbons of mown grass. These are stunning examples of efficient public transport systems seamlessly combined with attractive landscape design. Examples abound of landscape architects working with engineers to humanize transport infrastructure. In Lund, Sweden, landscape architect Sven-Ingvar Andersson (1927–2007) transformed a linear space along a railway line by laying sett paving and planting linden trees to create a pedestrian mall. On a larger scale, the Dutch practice West 8 has systematically planted thousands of birch trees in the spaces around and between Schiphol airport's runways and buildings—the species was chosen not just for the beauty of its bark, but because it is not attractive to perching birds and thus no threat to aeroplanes. In many instances, transport infrastructure can also become part of green infrastructure. Green transport corridors can be particularly valuable in linking together ecologically valuable patches of habitat within the urban matrix. Sometimes, when roads or railways would serve to divide, landscape architects are able to provide a remedy in the form of green bridges which connect habitats on either side.

The examples given in the last paragraph are cases of tactical interventions to turn elements of transport infrastructure into amenable places, but there is a broader, more strategic sense in which transport systems shape cities. A well-known example is 'Metroland', the swathe of suburbs that were built to the north-west of London in the early 20th century, served and facilitated by the Metropolitan Railway. The celebrated Copenhagen 'Finger Plan' of 1947 outlined a strategy whereby the city would be developed along five radiating commuter train lines (the 'fingers') extending from the dense urban centre (the 'palm'), but between these would lie wedges of greenspace for agriculture and recreation. As is so often the case, the pattern of the transport network, the built form of the city, and the structure of the open space system were intimately linked. Recognizing these linkages provides tools for urban planning. Despite design codes and zoning regulation, there is much about the growth of cities which must be left to the market, often to the chagrin of planners and urban designers. However, the expenditure of public money on infrastructure projects and on urban greenspace is politically accepted, even in the most capitalist economies, so here there is often scope to shape the city for the common good, or at least in those societies where infrastructure precedes development. In the case of the informal settlements which are so much part of the megacity, the situation is rather different, though public funds can be provided for the retrofitting of infrastructure, open space, and services in places which sprang up without them.

Landscape urbanism and ecological urbanism

Just as debates at Harvard preceded the formation of urban design as a discipline, a conference at the University of Illinois, Chicago, in 1997 promulgated a new *ism* called 'landscape urbanism', a name coined by Charles Waldheim, who is now Chair of the Department of Landscape Architecture at Harvard Graduate School of Design. In the words of James Corner, Professor of Landscape Architecture at the University of Pennsylvania, this new 'way of thinking and acting' has been made necessary by 'the failure of traditional urban design and planning to operate effectively in the contemporary city'. This sense of failure seems to stem from consideration of the way American cities have continued to sprawl horizontally, from astonishment at the speed with which megacities have grown in developing countries, and from the new phenomenon of the hollowing out of older industrial cities once the businesses that were their lifeblood close or relocate. This process is exemplified by Detroit, no longer the 'Motor City' but now an urban landscape represented by images of derelict factories and ruinous grand hotels. In all of these situations, the landscape urbanists argue, the urban planner is powerless, and the only thing left which can link a city together is its landscape. One way of conceptualizing this would be to say that the focus of attention has shifted from buildings as the basic blocks of the city to landscape as the glue or the medium which binds everything together.
In a way which parallels the emergence of urban design, landscape urbanists do not propose the formation of a new profession, but suggest that the conceptual fields of such disciplines as landscape architecture, civil engineering, urban planning, and architecture need to be integrated. Master's programmes in landscape urbanism sprang up at several North American universities and at the Architectural Association in London.

Just how innovative landscape urbanism might be and the extent to which it is a reworking of time-served ideas from the landscape architecture tradition have been topics of much discussion. One of the tenets of landscape urbanism is that it is more important how a landscape functions—what it does for us—than how it looks. This is very similar to the ideas expressed by advocates of green infrastructure planning, but as I have argued in this book, a concern for functionality was a notion present at the very beginning of landscape architecture, in the work of Olmsted and his successors. Landscape urbanists would probably agree, but where they controversially take issue with the Olmstedian tradition is with its advocacy of *rus in urbe*, the inclusion of Romanticized nature in the city. This is rejected as at best an irrelevance or at worst a kind of camouflage or deceit. They go further and argue that the way we speak about landscape on the one hand and cities on the other is conditioned through a '19th-century lens of difference and opposition'. They wish to argue that we should do away with the binary distinction between the urban and the rural. They would like us to recognize that the city's footprint extends well into what we would traditionally call the countryside, and that the latter is organized to provide resources for the city, whether food, drinking water, or energy. At the same time, voids within the city, such as those created by the demise of an industry or areas associated with essential items of infrastructure, are opened up to natural processes such as ecological succession. Ever since the advent of Deconstruction as a literary and philosophical movement, it has been fashionable in academic circles to attack binary oppositions, but I would argue that many binaries, including this one, are quite useful and that the consequence of abolishing the distinction between countryside and town would be to strengthen the tendency towards sprawl and put at risk cultural landscapes adjacent to cities. Sometimes the rhetoric of landscape urbanism favours 'going with the flow' even if that means our cities will become radically decentred, rhizome-like networks, spread wide across the landscape. Yet it was concern about 'ribbon development' that led in Britain to planning laws and green belts to contain urban expansion. Unbridled capitalism and unchecked sprawl do not have to hold sway.

Sometimes good city planning means redirecting, slowing, or stopping things from happening.

On the other hand, there are many ideas within landscape urbanism which have great merit. Landscape urbanists like to take the long view, recognizing that sites and cities develop over time. In Corner's writing there is an emphasis upon preparing 'fields for action' or 'stages for performances'—phrases which are vague enough to refer either to the physical works such as the clearance of derelict buildings or to more abstract activities such as assembling parcels of land from different ownerships, raising funding, gaining various permissions, and so on, in order to allow things to happen with some degree of spontaneity. In place of fixed masterplans, landscape urbanism extols a flexible indeterminacy. Landscape urbanists write in praise of the sorts of urban gardening and agriculture that have sprung up on vacant land in Detroit. There is also a championing of neglected places, the left-over land and interstices between motorways, pipelines, sewage farms, railway sidings, and landfills. One project often referenced is the Parc de la Trinitat in Barcelona (1993), a park and sports complex tucked inside a looping highway interchange by designers Enric Batlle and Joan Roig. Equally celebrated is the more recent High Line in New York City (2005–10), where Corner's practice, Field Operations, collaborated with architects Diller Scofidio + Renfro to transform an abandoned elevated freight railroad on Manhattan into a linear park incorporating planting inspired by the self-seeded vegetation which had colonized the structure during many years of disuse (Figure 11).

We could add to this list of virtues by mentioning landscape urbanism's interest in making positive use of waste materials. In his book *Drosscape* Alan Berger, Associate Professor of Urban Design and Landscape Architecture at Massachusetts Institute of Technology, argues that all cities produce waste, but that this is something which can be scraped, shaped, surfaced, and

11. James Corner's landscape architecture practice, Field Operations, collaborated with architects Diller Scofidio + Renfro to transform an abandoned elevated freight railroad on Manhattan into the popular High Line linear park (2005–10)

reprogrammed to fulfil socially and environmentally useful purposes. Berger has written that 'the challenge for designers is thus not to achieve drossless urbanisation but to integrate inevitable dross into more flexible aesthetic and design strategies'. Field Operations' long-term involvement in transforming the Fresh Kills Landfill (2001–40) into what will eventually become New York's largest park is thus acclaimed as a beacon of landscape urbanist practice. Conceivably it has been necessary to push such arguments about the usefulness of waste land harder in North America, where historically the land supply has not been limited, than in crowded Europe, where land is tight and traditions of land reclamation grew out of the need to deal with war-damaged cities after the Second World War.

Landscape urbanism has been a deliberate and useful provocation. To hear the landscape described as 'machinic', to talk about dismantling the boundaries between disciplines, to think at vast physical and temporal scales, to downgrade or even dismiss the importance of aesthetics... these moves and others have had their desired effect, stimulating shifts in practice, new ways of conceptualizing urban issues and new approaches to imagining solutions. It was never intended to replace landscape architecture; one can be a landscape architect and a landscape urbanist, indeed it is important that those who enter the nexus of landscape urbanism bring with them their own specialist knowledge and skills. But landscape urbanism's arc as the radical idea of the moment is almost complete. Charles Waldheim suggested in 2010 that landscape urbanism had entered a 'robust middle age' which was a bit of a surprise for those outside America who were only just encountering it. In 2009, Harvard University held another conference, this time on the theme of ecological urbanism, an expansion of the landscape urbanist idea led by Mohsen Mostafavi, Dean of the Graduate School of Design. Whether the world is ready for yet another *ism* before the sun has set on the last one is a moot point, but the newcomer has retained many of the ideas which informed its predecessor, including the need for the design disciplines to respond to the scale of the ecological crisis which confronts us all. It calls for new ways of planning future cities as well as retro-fitting existing ones, and it seems to have ditched some of the more strident and off-putting aspects of landscape urbanism, including its impenetrable jargon. It is clear, though, that the values and perspectives of landscape architecture will continue to be central to this new movement. Landscape architects have been ecological urbanists for a very long time.

References

Preface

European Landscape Convention, Council of Europe, 2000.

Chapter 1: Origins

Downing, Andrew Jackson, *A Treatise on the Theory and Practice of Landscape Gardening, Adapted to North America*, C.M. Saxton, 1841.

Hogarth, William, *The Analysis of Beauty, Written with a View of Fixing the Fluctuating Ideas of Taste*, W. Hogarth, 1753.

Chapter 3: Modernism

Steele, Fletcher, 'New Pioneering in Garden Design', *Landscape Architecture*, 20, no.3 (April 1930): 162.

Tunnard, Christopher, *Gardens in the Modern Landscape*, The Architectural Press, 1938.

Chapter 4: Use and beauty

Williams-Ellis, Clough, *England and the Octopus*, Geoffrey Bles, 1928.

Howard, Ebenezer, *Garden Cities of Tomorrow*, Swan Sonnenschein & Company, Limited, 1902, first published in 1898 as *Tomorrow, A Peaceful Path to Real Reform*.

Thayer, Robert, *Gray World, Green Heart: Technology, Nature and the Sustainable Landscape*, Wiley, 1997.

Chapter 5: An environmental discipline

Wordsworth, William, *A Guide Through the District of the Lakes*, fifth edition, 1835, first published as an introduction to Joseph Wilkinson's *Select Views in Cumberland, Westmoreland and Lancashire*, 1810.

Carson, Rachel, *Silent Spring*, Houghton Mifflin, 1962.

Thayer, Robert, *LifePlace: Bioregional Thought and Practice*, University of California Press, 2003.

Lyle, John Tilman, *Regenerative Design for Sustainable Development*, Wiley, 1994.

Chapter 6: The place of art

Ross, Stephanie, *What Gardens Mean*, University of Chicago Press, 1998.

Hall of Shame website maintained by the Project for Public Spaces <http://www.pps.org/great_public_spaces/list?type_id=2> (accessed 24.02.2014).

Chapter 7: Serving society

Mozingo, Louise A. and Jew, Linda L. (eds), *Women in Landscape Architecture: Essays on History and Practice*, McFarland & Co. Inc. Publishers, 2012.

Chapter 10: Landscape and urbanism

City Population (website) Population Statistics for Countries, Administrative Areas, Cities and Agglomerations, with Interactive Maps–Charts. <http://www.citypopulation.de/> (accessed 24.02.2014).

Berger, Alan, *Drosscape*, Princeton Architectural Press, 2007.

Further reading

There are some good visual histories of the designed landscape which might complement this *Very Short Introduction* where the space available for illustration has been necessarily limited. A perennial favourite is *The Landscape of Man: Shaping the Environment from Prehistory to the Present Day* (3rd edition, Thames & Hudson, 1995) written by Geoffrey Jellicoe, Britain's most eminent 20th-century landscape architect, and illustrated with his sketches and photographs by his wife, Susan Jellicoe. William Mann's *Landscape Architecture: An Illustrated History* covers the same ground with plans and drawings but no photographs. Another good historical survey is Tom Turner's *Garden History: Philosophy and Design* 2000 *BC*–2000 *AD* (Routledge, 2005).

For anyone thinking of studying to enter the profession, there are several good introductory textbooks. Tim Waterman's *The Fundamentals of Landscape Architecture* is concise, well-written, and well-illustrated (AVA Publishing, 2009). A much heftier book, at least in size, is Barry Starke and John Ormsbee Simonds' *Landscape Architecture: A Manual of Environmental Planning and Design*, which is now in its 5th edition (McGraw-Hill Professional, 2013). Catherine Dee's *To Design Landscape: Art, Nature & Utility* (Routledge, 2012) is very approachable and beautifully illustrated, as is her earlier book *Form & Fabric in Landscape Architecture: A Visual Introduction* (Taylor & Francis, 2001). My own *Ecology, Community and Delight* (E. & F. N. Spon, 1999) and *Rethinking Landscape* (Routledge, 2007) are concerned with the concepts and values that are inherent in landscape architectural practice. Susan Herrington has also probed these matters in *On Landscape*

(Routledge, 2008), which is part of the Thinking in Action series. There have been two credible attempts to collate landscape architectural theory: *Theory in Landscape Architecture: A Reader*, edited by Simon Swaffield (University of Pennsylvania Press, 2002), and *Landscape Architecture Theory: An Evolving Body of Thought* by Michael D. Murphy (Waveland Press, 2005).

There are numerous good biographies of particular landscape gardeners and landscape architects. In view of Frederick Law Olmsted's centrality in the transition from gardening to landscape architecture, I would recommend Witold Rybczynski's *A Clearing in the Distance: Frederick Law Olmsted and America in the 19th Century* (Prentice Hall and IBD, 2000). Janet Waymark's *Thomas Mawson: Life, Gardens and Landscapes* (Frances Lincoln, 2009) covers the career of the first president of Britain's Institute of Landscape Architects. Brenda Colvin's contribution to the discipline is presented in Trish Gibson's *Brenda Colvin: A Career in Landscape* (Frances Lincoln, 2011). Ian McHarg's *A Quest for Life: An Autobiography* (John Wiley & Sons, 1996) is characteristically entertaining. Similarly, Lawrence Halprin's colourful career emerges vividly from his autobiography *A Life Spent Changing Places* (University of Pennsylvania Press, 2011). There are, of course, numerous monographs presenting the work and ideas of particular designers or design firms, far too many to catalogue here. There are also, from time to time, large compendiums which present a wide range of current practice. A fairly recent one is *1000 x Landscape Architecture* (Braun, 2009). There is also Philip Jodidio's *Landscape Architecture Now!* (Taschen, 2012). These make good coffee table books and might also be good sources for design ideas.

Some landscape architects have been good writers as well as talented designers, so there is a corpus of classic books which I should mention. When first published, Thomas Church's *Gardens are for People* ushered in Modernist garden design (3rd revised edition, University of California Press, 1995). Garrett Eckbo's *Landscape for Living*, first published in 1950, is now back in print (University of Massachusetts Press, 2009). Ian McHarg's *Design with Nature* is often said to have been the most influential book ever published by a landscape architect, and the 25th anniversary edition (John Wiley, 1995) is still available.

For readers who wish to learn more about Modernism, a key book is Peter Walker's *Invisible Gardens: The Search for Modernism in the*

American Landscape (MIT Press, 1996). Marc Treib has also written two important books: *Modern Landscape Architecture: A Critical Review* (MIT Press, 1994) and *The Architecture of Landscape, 1940–1960* (University of Pennsylvania Press, 2002). Another good survey of Modernist work is Janet Waymark's *Modern Garden Design: Innovation Since 1900* (Thames & Hudson, 2005). The influence of Minimalism and Land Art can be explored in John Beardsley's, *Earthworks and Beyond* (4th revised edition, Abbeville Press, 2006) and in Jeffrey Kastner's *Land and Environmental Art* (Phaidon Press, 2010). For the career of a seminal figure who crossed disciplinary boundaries, see *The Life of Isamu Noguchi: Journey without Borders* (Princeton University Press, 2006) by Masayo Duus. The work of another significant, if often controversial, practitioner is presented in *Recycling Spaces: Curating Urban Evolution: The Landscape Design of Martha Schwartz Partners* by Emily Waugh (Thames & Hudson, 2012).

If you wish to read more deeply into the environmental aspect of landscape architecture, you could begin with Aldo Leopold's *A Sand County Almanac & Other Writings on Ecology and Conservation* (reprint edition, Library of America, 2013). Robert Thayer's *Gray World, Green Heart: Technology, Nature and the Sustainable Landscape* (new edition, John Wiley & Sons, 1997) considers the human relation to technology and the role landscape architects have sometimes played in disguising it. The same author's *LifePlace: Bioregional Thought and Practice* (University of California Press, 2003) is also worth reading. For an easy introduction to landscape ecology, I recommend *Landscape Ecology Principles in Landscape Architecture and Land-Use Planning* by Wenche Dramstad, James D. Olson, and Richard T. T. Forman (Island Press, 1996). John Tilman Lyle's *Regenerative Design for Sustainable Development* (John Wiley & Sons, 1996) is of similar vintage and still relevant.

The problems and potentials of brownfield sites are explored in a number of books, notably *Principles of Brownfield Regeneration: Cleanup, Design, and Reuse of Derelict Land* by Justin Hollander, Niall Kirkwood, and Julia Gold (Island Press, 2010) and *Manufactured Sites* by Niall Kirkwood (reprint, Taylor & Francis, 2011). Alan Berger's *Drosscape: Wasting Land in Urban America* (Princeton Architectural Press, 2007) has been controversial because it acknowledges the inevitability of waste and sprawl and even finds some beauty in it.

Reclaimed brownfields also feature in Julia Czerniak and George Hargreaves' *Large Parks* (Princeton Architectural Press, 2007), while the work of the German practice Latz + Partner, who have developed an influential approach to the redesign of post-industrial sites is explored in Udo Weilacher's *Syntax of Landscape: The Landscape Architecture of Peter Latz and Partners* (Birkhäuser, 2007).

For those interested in landscape planning, Tom Turner's *Landscape Planning and Environmental Impact Design* (2nd edition, Routledge, 1998) is still relevant, but also see Paul Selman's *Planning at the Landscape Scale* (Routledge, 2006) and *Sustainable Landscape Planning: The Reconnection Agenda* (Routledge, 2012). See also *Resilience and the Cultural Landscape: Understanding and Managing Change in Human-Shaped Environments*, edited by Tobias Plieninger and Claudia Bieling (Cambridge University Press, 2012). Many books are now being published about green infrastructure: see, for example, *Green Infrastructure: Linking Landscapes and Communities* by Mark Benedict and Edward McMahon (Island Press, 2006), and *Sustainable Infrastructure: The Guide to Green Engineering and Design* by S. Bry Sarte (John Wiley & Sons, 2010).

The problems of cities are explored from a variety of perspectives in *The City Reader*, edited by Richard LeGates and Frederic Stout (Routledge, 2012), which includes Sherry Arstein's seminal article 'A Ladder of Citizen Participation'. *Groundwork: Partnership for Action*, edited by Walter Menzies and Phil Barton. (CreateSpace Independent Publishing Platform, 2012), tells the story of the influential environmental charity which now employs many landscape architecture graduates in Britain.

The close relationship between landscape architecture and urban design is at the heart of *Basics Landscape Architecture 01: Urban Design* by Tim Waterman and Ed Wall (Ava Publishing, 2009) and in Jan Gehl's *Cities for People* (Island Press, 2010). Landscape urbanism's particular slant on the problems of the city is presented from a series of perspectives in *The Landscape Urbanism Reader* edited by Charles Waldheim (Princeton Architectural Press, 2006). Critics of landscape urbanism are given space in *Landscape Urbanism and Its Discontents: Dissimulating the Sustainable City*, edited by Andres Duany and Emily Talen (New Society Publishers, 2013). The likely successor to landscape urbanism is showcased in *Ecological Urbanism* edited by Mohsen Mostafavi and Gareth Doherty (Lars Muller Publishers, 2010). The much-lauded High Line project in New York has inspired a number of books including *High Line: The Inside Story of New York City's Park in the Sky* by Joshua David and Robert Hammond (Farrar Straus Giroux, 2011).

“牛津通识读本”已出书目

古典哲学的趣味
人生的意义
文学理论入门
大众经济学
历史之源
设计，无处不在
生活中的心理学
政治的历史与边界
哲学的思与惑
资本主义
美国总统制
海德格尔
我们时代的伦理学
卡夫卡是谁
考古学的过去与未来
天文学简史
社会学的意识
康德
尼采
亚里士多德的世界
西方艺术新论
全球化面面观
简明逻辑学
法哲学：价值与事实
政治哲学与幸福根基
选择理论
后殖民主义与世界格局

福柯
缤纷的语言学
达达和超现实主义
佛学概论
维特根斯坦与哲学
科学哲学
印度哲学祛魅
克尔凯郭尔
科学革命
广告
数学
叔本华
笛卡尔
基督教神学
犹太人与犹太教
现代日本
罗兰·巴特
马基雅维里
全球经济史
进化
性存在
量子理论
牛顿新传
国际移民
哈贝马斯
医学伦理
黑格尔

地球
记忆
法律
中国文学
托克维尔
休谟
分子
法国大革命
民族主义
科幻作品
罗素
美国政党与选举
美国最高法院
纪录片
大萧条与罗斯福新政
领导力
无神论
罗马共和国
美国国会
民主
英格兰文学
现代主义
网络
自闭症
德里达
浪漫主义
批判理论

德国文学
戏剧
腐败
医事法
癌症
植物
法语文学
微观经济学
湖泊
拜占庭

儿童心理学
时装
现代拉丁美洲文学
卢梭
隐私
电影音乐
抑郁症
传染病
希腊化时代
知识

电影
俄罗斯文学
古典文学
大数据
洛克
幸福
免疫系统
银行学
景观设计学